KEY SKILLS

APPLICATION OF NUMBER

Level 2

Jill Gask

ISBN 1 84224 322 5

GNVQ Key Skills Unit - Application of Number Level 2
Published by Liberty Hall Ltd

© Copyright Liberty Hall Ltd, 2005

ISBN 1 84224 322 5

A CIP catalogue record for this book is available from the British Library

First edition 2005

All rights reserved

Without limiting the rights under copyright reserved above, **no part of this publication may be photocopied** or reproduced, stored in or introduced into a retrieval system, or transmitted in any form or by any means, electronic, mechanical, photocopying, recording, or otherwise, without prior permission of the above publisher of this book.

Liberty Hall does its very best to ensure that its books are accurate and up-to-date. However, Liberty Hall gives no warranty as to the completeness of the information contained within this book and takes no responsibility for any loss or damage to the purchaser of this book caused by any of the book's contents.

Printed by Liberty Hall Ltd

Acknowledgements

Screen shots and icons reprinted with permission from Microsoft Corporation
Windows, Word, Access and Excel are trademarks of Microsoft Corporation in the U.S.A. and other countries.
Clipart produced by Corel Corporation and GSP

To my husband and children

The author

Jill Gask has worked in computing for over 30 years. She currently teaches in Further Education.

Liberty Hall Ltd is a publishing company producing books for students of IT, Computing and Business.

Liberty Hall believes that:

- Education is the key to freedom.
- Good education is achieved through good teaching.
- Good teaching is supported by sound educational material.
- Books should be designed for the students.
- Books should be affordable.

Liberty Hall's books are written by good, experienced teachers who share these values.

Contents

This Book   1

Chapter 1 - Using Spreadsheets

1.1   Introduction - What is a Spreadsheet?   5
1.2   Columns and Rows   6
1.3   Menus   6
1.4   Toolbars   6

Chapter 2 - Directed Numbers

2.1   Introduction   15
2.2   Addition and Subtraction   15
2.3   Multiplication   17
2.4   Division   17
2.5   Order of Calculation   18

Chapter 3 - Conversion of Units

3.1   Introduction   19
3.2   Degrees Centigrade and Fahrenheit   21
3.3   Foreign Exchange   21
3.4   Spreadsheet Worked Example 1   16

Chapter 4 - Fractions

4.1   Introduction to Fractions   25
4.2   Equivalent Fractions   26
4.3   Converting a Fraction to its Simplest Term   26
4.4   Converting an Improper Fraction to a Mixed Number   27
4.5   Converting a Mixed Number to an Improper Fraction   27

Chapter 5 - Ratio

5.1   Introduction to Ratio   29
5.2   Simplifying Ratios   29
5.3   Proportional Values   31

Chapter 6 - Maps, Scales and Plans

6.1   Introduction   37
6.2   Map Scales   37
6.3   Plans   42
6.4   Three-Dimensional Representation   44

Chapter 7 - Percentages

7.1	What is Meant by a Percentage?	45
7.2	Expressing Values as a Percentage	45
7.3	Finding a Percentage of a Quantity	45
7.4	Increasing an Amount by a Given Percentage	48
7.5	Decreasing an Amount by a Given Percentage	48

Chapter 8 - Presentation of Data

8.1	Introduction	53
8.2	Raw Data	53
8.3	Tally Chart	53
8.4	Frequency Distribution	54
8.5	Types of Variables - Continous and Discrete	55
8.6	Grouped Distributions	55
8.7	Pie Charts	58

Chapter 9 - Averaging Data

9.1	Introduction	61
9.2	Arithmetic Mean	61
9.3	Median	62
9.4	Mode	62
9.5	Comparisons and Benefits of Mean, Mode and Median	62
9.6	Range	63

Chapter 10 - Perimeters and Areas

10.1	Introduction	65
10.2	Quadrilaterals	65
10.3	Parellelogram	66
10.4	Area of a Rectangle	68
10.5	Area of a Parallelogram	70
10.6	Area of a Triangle	71
10.7	Circles	72
10.8	Area of a Circle	75

Chapter 11 - Volumes

11.1	Introduction	77
11.2	Cuboid	77
11.3	Cube	77
11.4	Cylinders	79

Chapter 12 - Assignments   81

Appendix - Answers   85

Index   91

This Book

This book has been designed to support students studying for a level 2 qualification. It provides learning material for Key Skills Unit Level 2 in Application of Number. Many students will achieve their key skills as part of their progress through their main programme of study. Others will need to have extra support to achieve the required standard. This book helps provide this extra support.

This unit provides the material for students to gain the necessary number skills for Level 2. It provides the learning material for each of these skills and activities for practising them. This will provide a basis for evidence of achievement.

This unit covers three key skill areas. These are:

Interpreting Information from Different Sources

Carrying Out Calculations, and

Interpreting Results amd Presenting Findings.

It requires that students acquire a range of skills in these areas and provide evidence of these skills, i.e. demonstrate these skills by showing that they can do what is required or by providing documents or other material evidence which show that they have acquired the necessary skills.

2.1 Interpreting Information

In interpreting information, you need to know how to:

obtain relevant information from different sources
(eg from written and graphical material, first-hand by measuring or observing);

read and understand graphs, tables, charts and diagrams

read and understand numbers used in different ways, including negative numbers
(eg for losses in trading, low temperatures);

estimate amounts and proportions;

read scales on a range of equipment to given levels of accuracy
(eg to the nearest 10mm or nearest inch);

make accurate observations *(eg count the number of customers per hour)*;

select appropriate methods for obtaining the results you need, including grouping data when this is appropriate *(eg heights, salary bands)*.

Evidence

N2.1
Interpret information from a suitable source.

You must show you can

2.1.1 choose how to get the information you need to meet the purpose of your activity

2.1.2 obtain relevant information

2.1.3 choose appropriate methods to get the results you need.

Examples of evidence

Obtaining information: from a health club leaflet about physical fitness and health statistics in a newspaper; from bus/train timetables and details of actual travel times over a period of time; from a small business, showing income and expenditure over three months.

2.2 Carrying Out Calculations

In carrying out calculations, you need to know how to:

carry out calculations involving two or more steps, with numbers of any size with and without a calculator;

show clearly your methods of carrying out calculations and give the level of accuracy of your results;

work with and convert between fractions, decimals and percentages;

convert measurements between systems
(e.g. from pounds to kilograms, between currencies);

work out areas and volumes
(e.g. area of an L-shaped room, number of containers to fill a given space);

work out dimensions from scale drawings
(e.g. using a 1:20 scale);

use proportion and calculate using ratios where appropriate;

compare sets of data of an appropriate size such as 20 items each
(e.g. using percentages, using mean, median, mode);

use range to describe the spread within sets of data;

understand and use given formulae
(e.g. for calculating volumes, areas such as circles, insurance premiums, V=IR for electricity);

check your methods in ways that pick up faults and make sure your results make sense.

Evidence

N2.2
Carry out calculations to do with:
- a amounts and sizes;
- b scales and proportion;
- c handling statistics;
- d using formulae.

You must show you can

2.2.1 carry out calculations, clearly showing your methods and levels of accuracy

2.2.2 check your methods to identify and correct any errors, and make sure your results make sense.

Examples of evidence

Calculations to: compare health statistics at national and local level and work out a fitness programme or healthy diet (scaling down or up to meet needs, using suitable formulae); compare mean differences, range and median between advertised travel times and actual travel times; scale up likely business profits and losses over the next six months of trading.

2.3 Interpreting Results and Presenting Findings

In interpreting results and presenting findings, you need to know how to:

select effective ways to present your findings;

construct and use tables, charts and graphs and label with titles, scales, axes, and keys as appropriate;

highlight the main points of your findings and describe your methods;

describe what your results tell you and how they meet your purpose.

Evidence

N2.3

Interpret the results of your calculations and present your findings.

You must show you can

2.3.1 select effective ways to present your findings

2.3.2 present your findings clearly using a chart, graph or diagram and describe your methods

2.3.3 use more than one way of presenting your findings

2.3.4 describe what your results tell you and how they meet your purpose.

Examples of Evidence

Presenting findings using:

a graph of local and national health statistics, a pie chart of items in a healthy diet and a diagram showing height/weight relationships;

a graph of actual and advertised travel times, a bar chart of one person's journey times and a network diagram of travel routes;

a time line graph showing results of scaling up likely profits and loss, a pie chart of main items of expenditure and a diagram to show plans for improving the business.

Chapter 1
Using Spreadsheets

Objectives

At the end of this chapter you will be able to

- describe a spreadsheet
- enter data into a spreadsheet

1.1 What is a Spreadsheet?

Spreadsheets are an important tool for use in all types of business calculations.

A spreadsheet is the electronic equivalent of a large sheet of paper, ruled in columns and rows and a calculator. It allows the user to store data, names and numbers and do calculations. A simple spreadsheet might be:

Salesperson	Sales		
	Oct	Nov	Dec
Brown, A	£2,300	£3,100	£1,900
Green, B	£3,500	£4,200	£3,600
Gray, J	£1,600	£2,000	£2,100
TOTAL	£7,400	£9,300	£7,600

You can see that the spreadsheet has names, numbers and calculated totals.

The spreadsheet package used in this book is Microsoft Excel 5.0. Other spreadsheet packages have similar features.

1.2 Columns and Rows

The spreadsheet comprises columns and rows. The **columns** are identified by letters, and the letters are displayed in a border at the top of the spreadsheet. The **rows** are identified by numbers, and the number relating to each row is displayed in the left border. A **cell** is the intersection of a column and row; a **cell reference** contains the 'address' of a particular cell, i.e. its column letter followed by its row number, e.g. A3, or C7.

Each cell may contain one of three types of data:

* text (labels), e.g. for column or row headings

* numbers, i.e. particular values which may be used in calculations

* formulae, which tell the spreadsheet to perform certain calculations, e.g. to calculate the total of a column of values, or to work out the percentage that one value is of another value. The formulae ensure that these totals or percentages are updated automatically when any values in them are changed.

1.3 Menus

The menu bar at the top of the screen contains the list of menu options. Each option lets you perform a particular function. To choose an option, you need to:

open the menu, by clicking on the chosen menu option, e.g. **File** or **Edit** and

select the appropriate option from the pull-down menu, by clicking on that option.

The Edit Menu

1.4 Toolbars

Under the menu bar, are two toolbars. Each one consists of a row of buttons that will carry out spreadsheet tasks when you click on the appropriate button. Sometimes using a button is a quick alternative to using a menu command, but sometimes there is no menu alternative. The two toolbars in Excel are:

Standard toolbar, which has buttons to let you open or close files, print, use the autosum to add a column or row of values etc.

Formatting toolbar, which has buttons to let you modify the appearance and alignment of data in your worksheet

Activity 1 - Spreadsheet Exercise

Some students at a college run their own tuck shop, where they sell a small selection of snacks, including chocolate bars. Boxes of each type of chocolate bar are bought from a wholesaler; the bars are sold individually to student customers.

The students keep details of the number of boxes of chocolate bars bought and sold. The figures for the first three months are:

Type of Chocolate bar	January	February	March
Chocco	2	4	7
Whizz	3	6	5
Yumm	4	8	3
Trifick	2	3	6

The students have asked you to store these details in a spreadsheet and make further calculations, including the total price paid, income from selling the chocolate and the profit.

Step 1 - Enter the data into the spreadsheet

1. Enter the title **Tuck Box Sales - Boxes sold** in cell B2

   This label is too long to fit entirely in cell B2 - it overflows into adjacent cells to the is still attached to cell B2. It is termed a long label.

2. Enter the labels for the months of January, February and March and the types of chocolate bars, as shown below. Labels normally align to the left of the cell.

3. Enter the number of boxes of each type of chocolate bars sold for the three individual months as shown below. Numbers normally align to the right of the cell.

	A	B	C	D	E	F	G	H
1								
2		Tuck Shop Sales - Boxes Sold						
3								
4		January	February	March				
5	Chocco	2	4	7				
6	Whizz	3	6	5				
7	Yumm	4	8	3				
8	Trifick	2	3	6				

Step 2 - Find the total number of boxes of each type sold

We want to find the total number of boxes of each type of chocolate bar sold, by entering a formula to add up the number of individual boxes.

1. Enter the label **Total** in the cell to the right of March.
   Go down one cell, and use a formula to sum the number of boxes of Chocco bars sold.

   In Excel,
   Enter =sum(
   Move the cell pointer to the cell Containing the number of Chocco boxes sold in January, hold down the mouse button, then drag across to the cell under March
   [Three cells are enclosed by a dotted line - Check that the formula is correct]
   Enter the right bracket) - the formula should read =SUM(B5..D5)
   Press Enter

2. Now, use a similar formula to sum the number of boxes of Whizz bars sold.

3. Repeat the process for boxes of Yumm and Trifick bars.
   The worksheet will look like the one below:

	A	B	C	D	E	F	G	H
1								
2		Tuck Shop Sales - Boxes Sold						
3								
4		January	February	March				
5	Chocco	2	4	7	13			
6	Whizz	3	6	5	14			
7	Yumm	4	8	3	15			
8	Trifick	2	3	6	11			

Step 3 - Find the total cost of the boxes bought

a) Enter the cost price per box

1. Enter the label Cost Price in the cell to the right of Total.

2. Underneath the label Cost Price, enter the cost price in £ of each box of chocolate bars:
   Chocco cost £12.00, Whizz cost £21.80, Yumm cost £15.50 and Trifick cost £14.20 per box.
   (Do not enter the £ sign.)

b) Creating 2 decimal places

Highlight all the cells containing cost prices, and format all these cells to display 2 decimal places for the pence part, as shown on the next page.

   In Excel, highlight all the cost price cells.
   Select Format from the Menu bar.   Click on Cells, then Number.
   Click on Number again and choose 2 decimal places.   Click on OK

c) Calculate how much was paid for each box

The amount paid for the boxes of each type of chocolate bar is equal to the number of boxes multiplied by the cost price per box.

1. Add the label Price Paid in the cell to the right of Cost Price.

   In Excel,
   Go to the cell underneath Price Paid
   Enter =
   Click on the total number of Chocco bars
   Enter * (to multiply)
   Click on the cost price of a box of
   Chocco bars
   Press Enter

2. Underneath this label, enter a formula to calculate the individual price paid, by multiplying the total number of boxes by the cost price.

3. Repeat the calculation for the other three types of chocolate bar.

d) Calculate the total amount paid

1. Enter the label Total in cell A9 underneath Trifick.

   In Excel,
   Enter =sum(
   Move the cell pointer to the cell containing the price paid for Chocco boxes (156.00),
   hold down the mouse button, then drag down to the price paid for Trifick (156.20)
   [Four cells are enclosed by a dotted line -Check that the formula is correct]
   Enter the right bracket) - the formula should read =SUM(G5..G8)
   Press Enter

2. Now find the total price paid, by using a formula to sum all the values in the Price Paid column.

3. Make sure that the price paid is displayed with 2 decimal places throughout. (See Step 3(b))

	A	B	C	D	E	F	G	H
1								
2		Tuck Shop Sales - Boxes Sold						
3								
4		January	February	March	Total	Cost Price	Price Paid	
5	Chocco	2	4	7	13	12.00	156.00	
6	Whizz	3	6	5	14	21.80	305.20	
7	Yumm	4	8	3	15	15.50	232.50	
8	Trifick	2	3	6	11	14.20	156.20	
9	Total						849.90	

Step 4 - Find the total income from the sale of the boxes

a) Enter the sale price per box

Although the chocolate bars are sold individually, the sale price per box is used, since the students know how much each bar is sold for and how many bars there are in each box.

1. Enter the label Sale Price in the cell to the right of Price Paid.

2. Underneath the label Sale Price, enter the sale price in £ of each full box of chocolate bars: Chocco sells for £16.00, Whizz sells for £28.80, Yumm sells for £24 and Trifick sells for £21.60 per box. (Do not enter the £ sign)

3. Make sure that the sale price is displayed with 2 decimal places throughout. (See Step 3(b))

b) Calculate the income for each type of chocolate bar

The income for each type of chocolate bar is equal to the number of boxes multiplied by the sale price per box.

1. Add the label Income in the cell to the right of Sale Price.

2. Underneath this label, enter a formula to calculate the individual income, by multiplying the total number of boxes by the sale price.

   In Excel,
   Go to the cell underneath Income
   Enter =
   Click on the total number of Chocco bars (13)
   Enter *   *(to multiply)*
   Click on the sale price of a box of Chocco bars (16.00)
   Press Enter

3. Repeat the calculation for the other three types of chocolate bar.

c) Calculate the total income

Now find the total income, by using a formula to sum all the values in the Income column.

In Excel,
Enter =sum(
Move the cell pointer to the cell containing the income for Chocco boxes (208.00), hold down the mouse button, then drag down to the income for Trifick (237.60)
[Four cells are enclosed by a dotted line - Check that the formula is correct]
Enter the right bracket) - the formula should read =SUM(I5..I8)
Press Enter

Make sure that the income is displayed with 2 decimal places throughout. (See Step 3(b))

	A	B	C	D	E	F	G	H
1								
2		Tuck Shop Sales - Boxes Sold						
3								
4		January	February	March	Total	Cost Price	Price Paid	
5	Chocco	2	4	7	13	12.00	156.00	
6	Whizz	3	6	5	14	21.80	305.20	
7	Yumm	4	8	3	15	15.50	232.50	
8	Trifick	2	3	6	11	14.20	156.20	
9	Total						849.90	

Step 5 - Find the total Profit from the sale of the boxes

The profit is equal to the income minus the cost of the chocolate bars. We will find the profit for each type of chocolate bar, and then the total profit.

a) Calculate the profit on individual types of chocolate bar

1. Enter the label Profit in the cell to the right of Income.

2. Underneath the label Profit, enter a formula which will subtract the price paid from the income for boxes of Chocco.

   In Excel,
   Enter =
   Click on the Income for
   Chocco (208.00)
   Enter -   (minus)
   Click on the Price paid for Chocco (156.00)
   Press Enter

3. Repeat the calculation for the other three types of chocolate bar.

b) Calculate the total profit

Now find the total profit, by using a formula to sum all the values in the Profit column.

   In Excel,
   Enter =sum(
   Move the cell pointer to the cell containing the profit for Chocco boxes (52.00),
   hold down the mouse button, then drag
   down to the profit for Trifick
   [Four cells are enclosed by a dotted line
   - Check that the formula is correct]
   Enter the right bracket) - the formula
   should read =SUM(J5..J8)
   Press Enter

Make sure that the profit is displayed with 2 decimal places throughout. (See Step 3(b))

	A	F	G	H	I	J	K
1							
2		Tuck Shop Sales - Boxes Sold					
3							
4		Cost Price	Price Paid	Sale Price	Income	Profit	
5	Chocco	12.00	156.00	16.00	208.00	52.00	
6	Whizz	21.80	305.20	28.80	403.20	98.00	
7	Yumm	15.50	232.50	24.00	360.00	127.50	
8	Trifick	14.20	156.20	21.60	237.60	81.40	
9	Total		849.90		1208.80	358.90	

Step 6 - Save your worksheet

Now save your worksheet, by using one of the Menu options.

In Excel,
Click on File
From the pull down menu for File,
choose Save As.
Enter an appropriate file name, e.g. Tshop1.   Click on OK

Step 7 - Print your worksheet

Print your worksheet, by clicking on the Print icon or using one of the Menu options.

In Excel,
Select Page Setup then Page.
Select Landscape then OK.
*Select **File** from the Menu bar, then Print*
Click on OK

Step 8 - Display the formulae which you have used

1. Display your formulae by using the appropriate option.

In Excel,
*Click on **Tools** from the Menu bar,*
then Options, then View,
then Formulas
Click on OK

2. Now print your worksheet again, so that you will have a hard copy of your formulae.

Step 9 - Making corrections to your Tuck Shop model

If your formulae have been entered correctly, then any changes that you make to your original data should be reflected by the formulae, since they will be re-calculated automatically.
Imagine that the data you used initially was not correct, so make the following changes:

the number of Chocco bars sold in January should be 4,
the number of Whizz bars sold in February should be 8,
the number of Yumm bars sold in January should be 6,
and in March should be 2
and the number of Trifick bars sold in March should be 7.

Has the Total of all the chocolate bars changed?
Have all the individual prices paid and income (and the totals) changed?
Have the profits changed?
They should all have changed.

1. Save this worksheet as Tshop2.

2. Print a copy of the new figures.

Step 10 - Changing your Tuck Shop model

One major advantage of using a worksheet is that you are able to test out various assumptions, or options by altering values in the worksheet, and noting the effects. These are called **What if?** calculations.

This means that you can make changes to your model which would reflect what would happen if these changes were really made. Such changes might be to find out what would happen if:

the cost price of the chocolate bars increased or decreased,
the sale price increased or decreased,
the numbers of boxes sold was changed and
what the sale price should be in order to create a particular profit.

Activity 2 - Spreadsheet Exercise using What if? Calculations

Use the worksheet which you have just created and saved as Tshop2, i.e. after the changes you made in Step 9. The students committee has decided that they are not satisfied with the profit made on each type of chocolate bar, and believe that the sale price of the chocolate bars should be changed. You are asked to change the sale price of the chocolate bars up or down until the figures satisfy the following conditions:

The profit on Chocco bars should be £96.00,
The profit on Whizz bars should be £91.20,
The profit on Yumm bars should be £152.00,
The profit on Trifick bars should be £78.60,

1. Save your worksheet as Tshop3.

2. Print out a copy of your worksheet.

Chapter 2
Directed Numbers

Objectives

At the end of this chapter you will be able to

- add, subtract, multiply and divide
- calculate using the correct order

2.1 Introduction

Numbers which have a + or - attached to them are called directed numbers, e.g. +3 or -5. There are a few simple rules about using directed numbers.

If we want to say that a temperature is 5°C below freezing point, we say that the temperature is -5°C; the freezing point is at 0°C. We can think of a thermometer, with plus and minus temperatures, or we can use a vertical number line to represent directed numbers.

```
 5 ─  Positive numbers              ↑
 4 ─
 3 ─  Temperatures above
 2 ─  freezing point            Add numbers
 1 ─
 0 ─  Freezing point
-1 ─
-2 ─  Negative numbers
-3 ─                           Subtract numbers
-4 ─  Temperatures below
-5 ─  freezing point            ↓
```

2.2 Addition and Subtraction

We can think of addition as moving **up** the number line, and subtraction as moving **down** the number line.

Numbers with the same sign

Example 1: Find the value of + 2 + 4

Start at zero; move up 2; then move up 4 more; you are now at 6 above zero.

Solution: $2 + 4 = +6$

Example 2: Find the value of -1 - 3

Solution: $\quad -1 - 3 = -4$

Start at zero; move down 1; then move down 3 more; you are now at 4 below zero.

Example 3: Find the value of -1 - 3 - 1

Solution: $\quad -1 - 3 - 1 = -5$

Start at zero; move down 1; then move down 3; now move down 1; you are now at 5 below zero.

Hint: When you add numbers with the same sign, sum the numbers together, then give the answer the same sign.

Numbers with different signs

Example 4: Find the value of 4 - 3

Solution $\quad 4 - 3 = +1$

Start at zero; move up 4; then move down 3; you are now at 1 above zero.
Sometimes you may see 4 - 3 written as 4 + (-3) or 4 - (+3)

Example 5: Find the value of -5 + 3 + 4

Solution: $\quad -5 + 3 + 4 = -5 + 7$
$\quad\quad\quad\quad\quad\quad\quad\quad = 2$

Start at zero; move down 5; then move up 3; now move up 4; you are now at 2 above zero.

Example 6: Find the value of 2 - 6 + 7

Solution: $\quad 2 - 6 + 7 = 9 - 6 \quad$ *adding the plus values together*
$\quad\quad\quad\quad\quad\quad\quad\quad = 3$

Start at zero; move up 2; then move down 6; now move up 7; you are now at 3 above zero.

Hint: When you add numbers with different signs, find the difference between the numbers, then give the answer the sign of the largest number.

Exercise 2.1

Write down the answers to the following, without using a calculator:

1. + 7 + 5   2. -2 -5   3. 5 - 12   4. 6 - 14
5. -3 + 7   6. -10 + 4   7. -4 - 2 - 5   8. 17 - 6 - 3
9. -2 - 5 + 6   10. -4 + 1 - 7   11. 6 - 2 + 1   12. 4 - 9 - 2
13. 3 - 6 + 2   14. 4 + 7 - 6   15. 5 - 9 + 2

2.3 Multiplication

From the previous section, we can see that $4 + 4 + 4 = 12$
i.e. $3 \times 4 = 12$

i.e. when **two positive numbers** are multiplied together, the **result is positive**.

Also, from the previous section, $-4 - 4 - 4 = -12$
i.e. $3 \times -4 = -12$

i.e., when a **positive** and a **negative** number are multiplied together, the **result is negative**.

Problem: To find the value of -3×-4:

Solution: We have just shown that
$$3 \times -4 = -12$$
But $-3 = -(+3)$
So $-3 \times -4 = -(+3) \times -4$
$$= -(-12)$$
$$= +12$$

i.e., when a **negative** number is multiplied by a **negative** number, then the **result is positive**.

2.4 Division

The rules for dealing with + and - signs for division are similar to those for multiplication. From the multiplication figures above we can see that:

a) $12 \div 4 = \dfrac{12}{4} = 3$

b) $-12 \div 3 = \dfrac{-12}{3} = -4$

c) $-12 \div -4 = \dfrac{-12}{-4} = +3$

d) $12 \div -3 = \dfrac{12}{-3} = -4$

The rules for multiplication and division can be shown as follows:

When you multiply or divide two numbers with the **same** sign, the result is always **plus**.

When you multiply or divide two numbers with **different** signs, the result is always **minus**.

	Multiply or Divide	
By	+	-
+	+	-
-	-	+

Hint: When two signs are next to each other (and no number between them), e.g. 4 + -2, then use the multiplication rule, i.e.

 $4 + -1 = 4 - 1 = 3$
or $6 - +5 = 6 - 5 = 1$
or $7 - -3 = 7 + 3 = 10$

Exercise 2.2

Write down the answers to the following, without using a calculator:

1. 3×-3
2. -5×2
3. $24 \div 3$
4. $21 \div -3$
5. 4×6
6. $-20 \div -4$
7. -5×-2
8. $24 \div -3$
9. 7×-2
10. $15 \div 3$
11. -6×-5
12. $-30 \div 3$
13. $-10 \div -2$
14. -4×5
15. -2×-6

2.5 Order of Calculation

The order of calculation is:
- **B** Brackets
- **O** Of
- **D** Division
- **M** Multiplication
- **A** Addition
- **S** Subtraction

This means that the contents of brackets must be evaluated before using division, and that division must be done before multiplication, which must be done before addition, and finally subtraction.

Using these rules, we can see that:

working out the division first a)	$5 + 2 \times 3 - 8 \div 4$	=	$5 + 2 \times 3 - 2$
working out the multiplication next		=	$5 + 6 - 2$
working out the addition next		=	$11 - 2$
subtraction last of all		=	9
brackets first b)	$4 \times (2 + 3) - 7 + 3$	=	$4 \times 5 - 7 + 3$
multiplication next		=	$20 - 7 + 3$
addition next		=	$23 - 7$
subtraction last		=	26
start contents of brackets first c)	$(4 \times 3 - 2) \div -5$	=	$(12 - 2) \div -5$
brackets contents evaluated now		=	$10 \div -5$
division now		=	-2

Exercise 2.3

Write down the answers to the following, without using a calculator:

1. -6×3
2. $7 - -4$
3. $2 \times 3 + 3 \times 5$
4. $-4 \div 2 - 6$
5. $-6 \times (-2 + 1)$
6. $2 - -4 - +3$
7. $-6 - 4$
8. $-8 + 3$
9. $7 + 4 \times 2 - 3$
10. $2 \times (4 + 2) \div 3$
11. $7 \times (-3 + 5)$
12. $5 + (8 - 2) \times 2$
13. $6 - 2 \times 4 + 7$
14. $(7 \times 2 + 1) \div 3$
15. $(-11 + 2) \div (-3 + 6)$
16. $\dfrac{-4 \times 3}{-2}$
17. $\dfrac{3 \times -4 + \times -6}{-3}$
18. $\dfrac{10 - 2 \times -7}{4}$

Chapter 3
Conversion of Units

Objectives

At the end of this chapter you will be able to

- [ ] convert degrees centigrade to fahrenheit
- [ ] convert currency

3.1 Introduction

In continental Europe and many parts of the world, distances, weights and liquid measures are measured in metric units. The United States uses the imperial system of units. The UK, whilst originally using the imperial system, is gradually changing over to the metric system, to keep in line with Europe.

Typical conversion factors include:

1 inch	2.54 cm
1 foot	30 cm
1 yard	0.914 m
1 mile	1.609 km
1 pound (lb)	0.45 kg
1 pint	0.47 litres
1 gallon	4.55 litres

Example 1:
On holiday in France, Jane decides to cook a meal which requires various ingredients, including 2 pounds of apples and a quarter of a pound of sugar. How much of each item in kilograms should be used?

The conversion from pounds to kg is   1 lb =   0.45 kg

Thus   2 lb =   2 × 0.45 kg

         =   0.9 kg

Also, $\frac{1}{4}$ lb = $\frac{1}{4}$ × 0.45 kg

         =   0.1125 kg

So Jane should buy 0.9 kg apples and 0.1125 kg sugar

Example 2:

Jane also needs to buy some meat, which is pre-packed, with the weight given in kg. She wants to buy about $\frac{3}{4}$ lb meat; she cannot decide whether to buy the pack labelled 0.322 kg or the pack labelled 0.476 kg. If she decides to buy the pack which is nearest in weight to $\frac{3}{4}$ lb, which pack should she choose?

The conversion from pounds to kg is   1 lb = 0.45 kg

i.e.   1 kg = $\frac{1}{0.45}$ lb

The pack labelled 0.322 kg weighs $\frac{0.322}{0.45}$ lb = 0.71556 lb

The pack labelled 0.476 kg weighs $\frac{0.476}{0.45}$ lb = 1.05778 lb

The required amount of meat is $\frac{3}{4}$ lb = 0.75 lb

Thus the pack weighing 0.322 kg = 0.71556 lb is the one she should choose because this is the nearest to 0.75lb.

Exercise 3.1

1. A corner shop pre-packs various grocery items and labels them in both imperial and metric weights. Convert the corresponding weights for each of the following:

   a) 0.88 lb of tomatoes to kg.
   b) 800 g of cheese to lb.
   c) 1.2 lb of sausages to kg.
   d) 1.7 kg of carrots to lb.

2. A motor cycle petrol tank holds 3.2 gallons of petrol. Find

   a) the equivalent amount of petrol in litres
   b) if petrol costs 62.9 p per litre, how much it would cost (to the nearest p) to fill the tank.

3. When travelling on a charter airline, the luggage allowance is 15 kg. If a man's case weighs 23 pounds, find:

   a) the weight of his case in kg
   b) the weight in kg that the case is under the allowance
   c) convert the weight in b) to pounds

4. A blanket measures 6 ft 6 in by 5 ft. Find the size of the blanket in cm.

5. The distance from Villefranche to St. Amandin is 23 km. What is the distance in miles?

6. The maximum speed allowed on a motorway is 70 miles per hour. What is the speed in km per hour?

7. A nursery quotes the final height and spread of a particular shrub as 6 ft and 9 ft 6 ins. Find the final height and spread in metres.

8. A car has its petrol consumption quoted as 47 miles per gallon.
   a) What is the equivalent petrol consumption in km per gallon?
   b) What is the equivalent petrol consumption in km per litre?

3.2 Degrees Centigrade and Fahrenheit

The metric system uses degrees Celsius (° C), whilst the imperial system uses degrees Fahrenheit (° F).
Using the Celsius scale of temperature, the freezing point of water is at 0° C and the boiling point is at 100° C, whilst using the Fahrenheit scale, the freezing point is at 32° F and the boiling point is at 212° F.

Conversion to degrees Fahrenheit from degrees Celsius

$$°F = \left(\frac{9}{5} \times °C\right) + 32$$

Conversion to degrees Celsius from degrees Fahrenheit

$$°C = \frac{5}{9} \times (°F - 32)$$

Example 3:
The 'normal' body temperature for a person is 98.6°F. What is the equivalent temperature in degrees Celsius?

$$°C = \frac{5}{9} \times (°F - 32)$$

$$\text{'Normal' temperature} = \frac{5}{9} \times (98.6 - 32)$$

$$= \frac{5}{9} \times 66.6$$

$$\text{'Normal' temperature} = 37°C$$

Example 4: The temperature at Biarritz one winter day was 12° C. Find the temperature in degrees Fahrenheit.

$$\text{Temperature in }°F = \left(\frac{9}{5} \times °C\right) + 32$$

$$= \left(\frac{9}{5} \times 12\right) + 32$$

$$= 21.6 + 32$$

$$\text{Temperature in }°F = 53.6°F$$

 Activity 3 - Spreadsheet Exercise

Set up a spreadsheet to convert temperatures from degrees Fahrenheit to Celsius; start by converting 60° F to degrees Celsius.

To do the conversion, we will enter a formula similar to the one above. When using formulae, take care with the order of calculation, as it is often necessary to put brackets round particular parts of the formula, to ensure that the calculation is made correctly.

Enter the data into the spreadsheet, using a separate cell for each heading, and then enter 60 in the cell under 'Fahrenheit' as follows:

	Fahrenheit	Celsius	
	60		

In the cell under 'Celsius', enter a formula to calculate the corresponding temperature in degrees Celsius.

In Excel,
Start the formula with =
Then enter   (5/9)*(
Use the mouse to click on the cell containing the Fahrenheit temperature (60)
Enter -32)
Press Enter

	Fahrenheit	Celsius	
	60	15.56	

Thus the equivalent temperature to 60° F is 15.56° C.

Activity 4 - Spreadsheet Exercise

1. Convert the following temperatures from degrees Fahrenheit to Celsius, by entering the following temperatures in the spreadsheet above, and using the formula conversion:

    a) 100°F   b) 120°F   c) 0°F
    d) 20°F   e) 50°F

2. Create a spreadsheet model to convert from degrees Celsius to Fahrenheit, and, by using suitable formulae, convert the following temperatures from degrees Celsius to Fahrenheit:

    a) 140°C   b) 180°C   c) -10°C
    d) 3°C   e) 40°C

3.3 Foreign Exchange

Each country has its own currency. In the UK we use the £ sterling. When we travel from one country to another, or buy goods from another country, we need to exchange money into the currency of that foreign country. The exchange rate lets us convert from one country's monetary system to another.

The exchange rates for January 2000 were:

Country	Exchange rate for £1
France	10.4 French francs
Germany	3.11 Marks
Italy	3080 Lira
Spain	265 Pesetas
USA	1.55 Dollars

Example 5:
Robert plans to change £20 into Marks when he goes on holiday to Germany. How many Marks will he receive?

For every pound that Robert changes, he will receive 3.11 Marks
If he changes £20, he will receive 20×3.11 Marks
$$= 62.2 \text{ Marks}$$

Example 6:
Rachel bought a T-shirt in the USA for $8.71. What is the equivalent price in £ sterling?

Every $1.55 that Jenny spends is equivalent to £1
A T-shirt costing $8.71 will be worth $\quad 8.71 \div 1.55$ £
$\qquad$ Price in £ $\qquad = £5.62$

Exercise 3.2

1. Tim works for a travel agent and helps customers with their foreign currency. Find how much the following amounts will be when they are converted into the chosen currency.

   a) £20 into French francs
   b) £70 into U.S. dollars
   c) £150 into Spanish pesetas
   d) £500 into Italian Lira
   e) £55 into German marks

2. Change the following foreign currency into £ sterling to the nearest p.

   a) 500,000 lira
   c) 900 pesetas
   d) 25 dollars
   e) 36.7 marks

3. James went to Spain on holiday, and converted £80 into pesetas. He bought his brother a present costing 1160 pesetas and he bought his sister a present costing 1080 pesetas; he spent 9130 pesetas on himself.

   a) How many pesetas did he have initially?
   b) How many pesetas did he have left?
   c) What was this worth in £ sterling?

4. Julia went camping in Germany, France and Italy, and took £30 with her. She spent 17.4 Marks in Germany, 85 francs in France and 24562 lira in Italy. How much money in £ sterling did she have left?

5. Teresa took £60 with her when she went to France, and changed it all into francs. She spent 120 francs on rail travel, 87 francs on food and drink, 30 francs on souvenirs and 25 francs to visit a museum.

   a) How many francs did she have initially?
   b) How many francs did she have left?
   c) What was this worth in £ sterling?

Chapter 4
Fractions

At the end of this chapter, you will be able to

- [ ] use fractions
- [ ] convert fractions
- [ ] add and subtract fractions
- [ ] multiply fractions

4.1 Introduction to Fractions

In fractions, the whole of something is divided into a number of equal parts, e.g. if a bar of chocolate is divided into 8 equal parts, then each part represents one-eighth (written as 1/8) of that chocolate bar. The number below the line tells us how many equal parts there are, and is called the **denominator**. The number above the line tells us how many of these equal parts are taken and is called the **numerator**.

$$\text{Therefore, Fraction} = \frac{\text{Numerator}}{\text{Denominator}}$$

If we divide up rectangles, we can see that:

one divided by two   = $\frac{1}{2}$

one divided by three = $\frac{1}{3}$

one divided by four = $\frac{1}{4}$

i.e. the shaded part represents the fraction

If we shade several rectangles then we can show the value of any fraction, e.g.

3 out of the 5 squares are shaded,
i.e. the shaded part represents the fraction $\frac{3}{5}$

4 out of the 7 squares are shaded,
i.e. the shaded part represents the fraction $\frac{4}{7}$

4.2 Equivalent Fractions

There are many equivalent forms for each fraction. A pizza can be equally divided in many ways, and one-half of the pizza can be represented in many ways, e.g.

$\frac{6}{12}, \frac{4}{8}$ and $\frac{1}{2}$ are **equivalent fractions** because they all represent the same value.

To change $\frac{1}{2}$ to $\frac{4}{8}$ multiply both top and bottom by 4;

to change $\frac{1}{2}$ to $\frac{6}{12}$ multiply both top and bottom by 6.

The fraction is still one half of the pizza.

Example 1:

Complete $\frac{4}{7} = \frac{?}{35}$ to give the equivalent fraction.

Solution:
Comparing the denominators, 7 is increased to 35 by multiplying by 5
To get the equivalent value of the numerator, we must multiply 4 by 5

Therefore $\frac{4}{7} = \frac{4 \times 5}{7 \times 5} = \frac{20}{35}$

4.3 Converting a Fraction to its Simplest Terms

A fraction is in its simplest terms when both the numerator and denominator are whole numbers, and they cannot both be divided by the same number. For example, the fraction $\frac{8}{12}$ is not in its simplest terms (since both parts are divisible by 4), but $\frac{2}{3}$ is.

Example 2:
Show the fraction $\frac{42}{56}$ in its simplest terms.

Solution:
42 is divisible by 2, 3 and 7
56 is divisible by 2, 2, 2 and 7
So, both 42 and 56 are divisible by 2 and 7, i.e. are divisible by 14

$$\frac{42}{56} = \frac{14 \times 3}{14 \times 4} = \frac{3}{4}$$

4.4 Converting an Improper Fraction to a Mixed Number

Improper fractions are fractions which are 'top heavy', i.e. the numerator (value on the top) is greater than the denominator (the value below the line); this means that the value of the fraction is greater than 1, and that it will have a whole number and a fractional part, i.e. it is a **mixed number.**

Example 3:
Express $\frac{17}{3}$ as a mixed number.

Solution:
Divide 17 by 3, and note the remainder

$$17 \div 3 = 5 \quad \text{remainder } 2$$

$$\frac{17}{3} = 5\frac{2}{3}$$

4.5 Converting a Mixed Number to an Improper Fraction

To convert from a mixed number to an improper fraction, we need to convert the whole number into fractional parts.

Example 4:
Convert $4\frac{3}{5}$ to an improper fraction.

Solution: First, convert the whole number, 4, into fifths, and then add it to $\frac{3}{5}$

$$4\frac{3}{5} = \left(4 \times \frac{5}{5}\right) + \frac{3}{5}$$
$$= \frac{20}{5} + \frac{3}{5}$$
$$4\frac{3}{5} = \frac{23}{5}$$

Exercise 4.1

1. Complete the following to give equivalent fractions:

   a) $\frac{1}{2} = \frac{?}{4}$
   b) $\frac{10}{16} = \frac{?}{8}$
   c) $\frac{21}{60} = \frac{7}{?}$

   d) $\frac{3}{8} = \frac{12}{?}$
   e) $\frac{13}{18} = \frac{52}{?}$
   f) $\frac{20}{36} = \frac{?}{9}$

2. Write down the following fractions with the denominator (bottom number) stated:

   a) $\frac{2}{3}$ with denominator 12
   b) $\frac{3}{7}$ with denominator 21

   c) $\frac{3}{8}$ with denominator 40
   d) $\frac{2}{9}$ with denominator 72

   e) $\frac{4}{5}$ with denominator 60
   f) $\frac{2}{3}$ with denominator 48

3. Write down the following fractions in their simplest terms:

   a) $\frac{6}{12}$ b) $\frac{20}{25}$ c) $\frac{21}{63}$ d) $\frac{9}{15}$

   e) $\frac{35}{42}$ f) $\frac{18}{27}$ g) $\frac{210}{315}$ h) $\frac{85}{100}$

4. Write down the following fractions as mixed numbers:

   a) $\frac{4}{3}$ b) $\frac{13}{8}$ c) $\frac{17}{4}$ d) $\frac{14}{5}$

   e) $\frac{23}{7}$ f) $\frac{39}{9}$ g) $\frac{37}{10}$ h) $\frac{18}{4}$

6. Write down each of the following as improper (top heavy) fractions:

   a) $1\frac{2}{3}$ b) $3\frac{3}{4}$ c) $5\frac{2}{7}$ d) $7\frac{3}{20}$

   e) $2\frac{7}{9}$ f) $4\frac{1}{2}$ g) $2\frac{4}{5}$ h) $3\frac{5}{6}$

6. Adele works 8 hours a day in a café; during this time, she prepares food for 3 hours and waits at tables for 1½ hours. Write down as a fraction of her working day in its simplest terms, the time she spent:

   a) preparing food   b) waiting at table.

7. A company records that 280 printers were sold last week, and 35 of those were colour printers. What fraction of printers sold, in simplest terms, were colour printers?

8. In a class of 27 students, 6 are girls. What fraction of students, in simplest terms, are girls?

9. In a class of 28 students, 4 students travel to college by train, 7 travel to college by bus, 3 cycle and the rest of the students walk. What fraction of students, in simplest terms:

   a) Travel by bus   b) Travel by train
   c) Cycle to college   d) Walk to college

10. The dried fruit used in a cake recipe comprises 200 g raisins, 300 g currants, 250 g sultanas and 50g mixed peel.

a) Find the total weight of the dried fruit.

What fraction of the total amount of the dried fruit, in simplest terms, was:

b) Raisins   c) Currants
d) Sultanas   e) Mixed peel

Chapter 5
Ratio

At the end of this chapter you will be be able to

- simplify ratios
- use proportional values
- create scale diagrams and maps

5.1 Introduction to Ratio

A ratio compares one number with another. Ratios are used where the proportion of one part to another is constant. For instance, to make concrete, we mix sand and cement in the ratio 3:1 and then add water. This ratio is constant whether we measure the sand and cement in eggcups or any other (usually larger) type of container.

Similarly, if we make pastry, the ratio between the flour and fat is constant, whether we make enough for a single pie base or mix a larger amount to make a large number of pies.

The amount of the parts (or ingredients) must be in the **same ratio** no matter what volume is being used.

5.2 Simplifying Ratios

The ratio of fat to flour used to make pastry is 1:2. It would be difficult to make pastry with just 1 gram of fat and 2 grams of flour, but the ratio between the two items must be the same. This could be

    100g : 200g   or   250g : 500g   or   500g : 1000g
    fat : flour   or   fat : flour   or   fat : flour

The ratio 1 : 2 (and the ratio 3 : 1 for making concrete) is in its **simplest terms;** they are whole numbers and cannot both be divided further by the same number and remain whole numbers.

The ratio 100:200 is not in its simplest form, because both numbers can be divided by 100, i.e. 100 is a **common factor** of both 100 and 200.

Sometimes the common factor is not as obvious as it is in this case.

Example 1:
Show the ratio 15 : 21 in its simplest form.

Solution: 3 can be divided into both 15 and 21
$$15 \div 3 = 5$$
$$21 \div 3 = 7$$
Thus 5 : 7 are the simplest terms for the ratio 15 : 21

Example 2:
In a store selling PCs, an ordinary bubble jet printer costs £168 and a colour model costs £312. Express these prices as a ratio in its simplest terms.

Solution: The ratio between the printer prices is 168 : 312

The ratio can be simplified only if both values can be divided by the same common factor. Both values can be divided by 2, and this simplifies the ratio to

$$84 : 156$$

The factors of 84 are 3, 4 and 7; the factors of 156 are 3, 4 and 13, i.e. 3 and 4 are factors of both values. Dividing both values by 3, the ratio becomes

$$28 : 52$$

Dividing both values by 4, the ratio becomes

$$7 : 13$$

i.e. the simplest terms of the ratio of £168 : £312 are 7 : 13.

Exercise 5.1

1. Write the following ratios in their simplest terms:
   a) 3 : 18   b) 12 : 60   c) 14 : 35   d) 50 : 75   e) 150 : 195

2. Write the ratios for the following quantities in their simplest terms:
   a) 22 cm, 77 cm   b) 27 ml, 63 ml   c) 500 g, 2.5 kg
   d) 20 cm, 140 cm   e) 30 mm, 15 cm

3. In a class, there are 8 boys and 12 girls. What is the ratio of boys to girls in simplest terms?

4. A computer screen is 28 cm wide and 21 cm high. What is the ratio of width to height of the screen in simplest terms?

5. Amy has £30 and Bill has £12. What is the ratio in simplest terms of these amounts?

Exercise 5.1 (cont)

6. Complete the following ratios:
   a) $2:3 = 4:?$ b) $22:11 = ?:1$ c) $4:5 = 20:?$
   d) $?:5 = 27:45$ e) $7:? = 28:48$ f) $21:27 = 7:?$

7. The speed quoted for three processors is 90 Hz, 120 Hz and 150 Hz. Express these speeds as ratios in their simplest terms.

8. A box of tissues contains 150 tissues, and a handypack contains 12 tissues. Write the ratio of these quantities in their simplest terms.

5.3 Proportional Values

Often we need to know the original size of each part, given that we know the ratio and the final amount.

Example 3:
£240 is split between 3 people in the ratio $3:4:5$. How much does each person receive?

Solution: Firstly, we need to find the total number of parts which make up the whole amount.

    Total number of parts $= 3 + 4 + 5$
          $= 12$

Next, find how much each part represents.

    Value of each part $= £240 \div 12$
          $= £20$

Now, find out how much each person receives, by multiplying the number of parts a person has by the value of each part, i.e.

    First person receives $3 \times £20$ $= £60$
    Second person receives $4 \times £20$ $= £80$
    Third person receives $5 \times £20$ $= £100$

Check: Total amount $= £60 + £80 + £100 = £240$

Example 4:
The ratio of girls to boys taking a particular college course is 8 : 5. If the number of girls is 48, find

a) How many boys are there?
b) How many students are taking the course?

Solution:
a) The ratio is 8 girls : 5 boys, i.e. the number of girls is represented by 8 parts
   i.e. 8 parts represents 48 students
   1 part is represented by 48 ÷ 8 students
                       = 6 students

But the number of boys is represented by 5 parts, so
   Number of boys = 5 × 6 students
                 = 30 students

b) The total number of students = 48 + 30
                                = 78 students

Exercise 5.2

1. Divide £720 in the ratio 5 : 3.

2. Divide 120 kg in the ratio 2 : 3 : 5.

3. Divide £450 in the ratio 3 : 2 : 5 : 5.

4. Find the smallest share when £180 is divided in the ratio 6 : 5 : 2 : 7.

5. Anne and Ellie buy a box of floppy discs for £6.90. If they share the discs in the ratio 7 : 3, how much should each one pay?

6. £40 is divided between Steven and Jeremy in the ratio 2 : 3. How much more does Jeremy get than Steven?

7. A sum of money is divided between Suzie and Lucy in the ratio 2 : 5. If Suzie receives £7, how much does Lucy receive? What is the total sum of money?

8. A sum of money is divided in the ratio 7 : 15. If the smaller amount is £21, find the larger amount.

9. Find the weight of each share when a bag of sweets weighing 520 g is divided in the ratio 4 : 7 : 2.

10. £81 is shared between Ken and Luke in the ratio 2 : 7. How much more does Luke receive than Ken?

11. Alex and Chris share a bag of sweets in the ratio 4 : 7. If Alex has 32 g of sweets, how much does Chris receive?

Exercise 5.2 (cont)

12. The ratio of goals scored by United to Rovers is 7 : 12. If United has scored 56 goals:
    a) How many goals have Rovers scored?
    b) What is the difference in goals scored by the two teams?

Spreadsheet Worked Example 1: A Spreadsheet Model

Rose, Sam and Tom invest £3000, £5000 and £7000 in a business. The profit at the end of the first year is £4600. If the profit is split in the ratio of their investments, how much does each receive?

Step 1 - Enter the data into the spreadsheet

Enter the data into the spreadsheet, using a separate cell for each person's name and a separate cell for each amount invested, as shown below:

	Rose	Sam	Tom	
	2700	5000	7000	

Enter the label 'Total' in the cell to the right of 'Tom'
In the cell below the label 'Total', enter a formula to add the three investments.

In Excel,
Enter =sum(
Move the cell pointer to the cell containing Rose's investment (2700),
hold down the mouse button, then drag across to the cell under Tom
[Three cells are enclosed by a dotted line - Check that the formula is correct]
Enter the right bracket)   Press Enter

	Rose	Sam	Tom	Total	
	2700	5000	7000	14700	

Step 2 - Find the profit per share

Leave one row blank
On the next row, enter 'Profit at end of year'
On this row, underneath 'Total', enter the profit figure, i.e. 4600
Leave one row blank
On the next row, enter 'Profit per share'
On this row, underneath 'Total', enter a formula to calculate the profit per share; the profit per share is found by dividing the profit by the total number of shares

Spreadsheet Worked Example 1(cont)

In Excel,
Start the formula with =
Use the mouse to click on the cell containing the profit figure (4600)
Enter /(to divide)
Use the mouse to click on the cell containing the total shares (14700)
Press Enter

	Rose	Sam	Tom	Total	
	2700	5000	7000	14700	
	Profit at end of year			4600	
	Profit per share			0.31293	

The Profit per share is 0.31293.

Step 3 - Find how much each person receives

Since Rose invested £2700, she will receive 2700 of these 0.31293 shares, i.e. she will receive 2700 × 0.31293. Similarly, Sam will receive 5000 of these shares and Tom 7000 of these shares.

On your spreadsheet, leave one row blank
On the next row, enter 'Each receives'
On the next row, under the cell labelled 'Rose', enter a formula to calculate the amount which Rose receives.

In Excel:
Start the formula with =
Use the mouse to click on the cell containing Rose's shares (2700)
Enter * (to multiply)
Use the mouse to click on the cell containing the profit per share (0.31293)
Press Enter

Under the cell labelled 'Sam', enter a formula to calculate the amount Sam receives
Under the cell labelled 'Tom', enter a formula to calculate the amount Tom receives
Finally, under the cell labelled 'Total', enter a formula to add together the amounts received by Rose, Sam and Tom. *- See Step 1 above.*

	Rose	Sam	Tom	Total
	2700	5000	7000	14700
Profit at end of year			4600
Profit per share			0.31293
Each receives			
	844.90	1564.63	2190.47	4600

Check: Is this total profit the same as the profit earned at the end of the first year? It should be! If not, check you calculations!
Now, save your worksheet with the filename ratio1.
Print a copy of your worksheet.

Step 4 - Display the formulae which you have used

Now display your formulae used in your calculations.

In Excel, select Tools, then Options, then View, then click on Formulas.

Your spreadsheet should look like this, with formulae in the shaded cells.

	Rose	Sam	Tom	Total
	2700	5000	7000	=SUM(B3:D3)
Profit at end of year			4600
Profit per share			=E5/E3
Each receives			
	=B3*E7	=C3*E7	=D3*E7	=SUM(B10:D10)

The symbols in your spreadsheet should be the same as in the shaded cells on the previous page, although the individual cell references may be different.

Step 5 - Change the profit in your spreadsheet model

After further examination of the business figures in the above example, it was found that the profit at the end of the first year was £5700. Using this figure for profit, how much did Rose, Sam and Tom receive?

(When you change the profit in your worksheet, it should change to show the new amount that each received, **provided that you have used formulae in your spreadsheet model.**)

Step 6 - More changes to your spreadsheet model

Rose received a win on the lottery, and was able to invest £3200 (instead of £2700). If the first year profit was £5100, how much did each person receive?

Modify your spreadsheet with these new figures, and your figures should change to reflect these changes.

Activity 5 - Spreadsheet Exercises
Creating a New Spreadsheet Model

Create new spreadsheet models for each of the following:

1. Alan, Bill, Carol and Diane invest £5000, £12000, £16000 and £10500 respectively in a business. If profits of £9000 are shared between them in the ratio of their investments, how much does each person receive?

2. Three vineyard owners grow 25, 37 and 44 acres of vines respectively and decide to pool their crop for processing together. If the grapes are sold for £85,000, how much does each vineyard owner receive?

3. 5 people in an office jointly enter £5, £3, £6, £10 and £2 into the lottery each week. One week they won £105,463 and divided their winnings in the ratio of each person's contribution. How much did each person receive?

Chapter 6
Maps, Scales and Plans

At the end of this chapter you will be be able to

- describe scale diagrams and maps
- describe and draw plans
- describe and represent three-dimensionsl objects

6.1 Introduction

Scale diagrams and maps represent the original dimensions of an object; all measurements of the original and the map or scale diagram are in the same proportion. The scale on a map shows the proportion by which each measurement has been reduced, and this scale is usually stated as a ratio.

6.2 Map Scales

When a map is drawn to a scale of 1 : 20000, this means that a length of 1 cm on a map represents 20000 cm in real life. We would need to convert this distance to metres or kilometres, i.e.

Since 100 cm = 1 m, divide by 100 to convert from centimetres to metres (cm to m).
Since 1000 m = 1 km, divide by 1000 to convert from metres to kilometres (m to km).

$$
\begin{aligned}
20000\,\text{cm} &= 20000 \div 100\,\text{m} \\
&= 200\,\text{m} \\
&= 200 \div 1000\,\text{km} \\
&= 0.2\,\text{km}
\end{aligned}
$$

Example 1: A map is drawn to a scale of 1 : 50000. Calculate:

a) The length of a road which appears as 3 cm on the map.
b) The length on the map of a lake which is 7 km long in real life.

a) 1 cm on the map represents 50000 cm in real life
   Hence, 3 cm represents 3 × 50000 cm
   = 150000 cm
   = 150000 ÷ 100 m = 1500 m
   = 1500 ÷ 1000 km
   Real length of road = 1.5 km

b) 50000 cm in real life is represented by 1 cm on the map
   i.e. 1 cm in real life = 1 ÷ 50000 cm on the map
   But   7 km = 7 × 1000 m   = 7000 m
                  = 7000 × 100 cm = 700,000 cm
   Hence   7 km = 700,000 ÷ 50000 cm on the map
   7 km long lake = 14 cm long on the map

Exercise 6.1

1. On a map with a scale of 1 : 15000, the distance between two towns is 27.8 cm. What is the actual distance in km?

2. The scale of a map is 1 : 20000. On the map, a lake is 3.4 cm. What is the real length of the lake in km?

3. The distance between two stations is 32.5 km. How far apart are the stations on a map if the scale is 1 : 50000?

4. A map has a scale of 1 cm = 2 km.

   a) Express this scale as a ratio.
   b) What distance will 6.2 cm on the map represent?
   c) What distance will 4.8cm on the map represent?

5. If the scale of a map is 1 : 25000, what is the length on the map of a road which is 6 km long in real life?

6. The distance between two churches is 23 km. What distance will this be on a map which has a scale of 1 : 80000?

Spreadsheet Worked Example 2

You have been planning a walking holiday, and have used a number of maps; the scale used is different on each of the maps. You know the distance on the map and the scale of the map. You are asked to build up a spreadsheet model to use these figures, and to calculate the real distance, in cm, m, and km. The map distance and scales are:

Walk Number	1	2	3	4	5
Map Distance(cm)	2.9	7.8	4.2	10.5	12.3
Scale	1:20000	1:5000	1:15000	1:25000	1:8000

Step 1 - Enter the data into the spreadsheet

Create your spreadsheet, using suitable labels and enter the figures for Walk number 1 (leaving out the '1 :' in the scale), as shown below:

Walk Number	Distance on Map (cm)	Scale Factor	Real Distance (cm)	Real Distance (m)	Real Distance (km)	
1	2.9	20000				

Step 2 - Calculate the real distance

a) Real distance measured in cm

The real distance in cm is the distance on the map multiplied by the scale factor.

On the same row, but under the 'Real Distance (cm)' label, enter a formula to multiply the distance on the map (2.9) by the scale (20000).

In Excel,
Start the formula with =
Use the mouse to click on the map distance (2.9)
Enter *   *(to multiply)*
Use the mouse to click on the scale (20000)
Press Enter

Your spreadsheet should look like this:

Walk Number	Distance on Map (cm)	Scale Factor	Real Distance (cm)	Real Distance (m)	Real Distance (km)	
1	2.9	20000	58000			

It is difficult to imagine a distance of 58000 cm - we need to convert this to m or km.

In Excel, start the formula with =
Use the mouse to click on the real distance in cm (58000)
Enter /100
Press Enter

b) Real distance measured in m

Now calculate the Real Distance in metres: how do you convert from centimetres to metres? Divide the distance in cm by 100!
On the same row, but under the 'Real Distance (m)' label, enter a *formula* to divide the real distance in cm by 100

c) Real distance measured in km

Finally, calculate the Real Distance in kilometres: how do you convert from metres to kilometres? Divide the distance in m by 1000!
On the same row, but under the 'Real Distance (km)' label, enter a *formula* to divide the real distance in m by 1000.

In Excel, start the formula with =
Use the mouse to click on the real distance in m (580)
Enter /1000
Press Enter

Your spreadsheet should look like this:

Walk Number	Distance on Map (cm)	Scale Factor	Real Distance (cm)	Real Distance (m)	Real Distance (km)
1	2.9	20000	58000	580	0.58

Now repeat the calculations for the rest of the figures in the table.

Save your worksheet with the filename walk.
Print out your worksheet.

Spreadsheet Worked Example 3

You are on a bicycle touring holiday, and have travelled over a number of roads, tracks and bridleways. You know, from checking the odometer on your bike, the distances along each of these different parts. You have used a number of maps, and each one has a different scale. You are asked to build up a spreadsheet model to use these figures, and, using the real distances (in km) and the scale factor, to calculate the map distance in cm. The real distance (in km) and scales are:

Ride Number	1	2	3	4	5
Real Distance(km)	3.5	12.7	9.8	2.3	5.7
Scale	1:40000	1:25000	1:50000	1:20000	1:10000

Step 1 - Enter the data into the spreadsheet

Create your spreadsheet, using suitable labels and enter the figures for Ride number 1 (leaving out the '1 :' in the scale), as shown below:

Ride Number	Real Distance (km)	Scale Factor	Real Distance (m)	Real Distance (cm)	Distance on Map (cm)
1	3.5	40000			

Step 2 - Convert real distance units

To find the map distance in cm, we need to convert the real distance to cm, and then divide by the scale factor. We will do this by firstly converting the real distance to metres, and then to cm, before dividing by the scale factor.

a) Real distance in m

On the same row, but under the 'Real Distance (m)' label, enter a formula to multiply the real distance on the map (3.5) by 1000 (since 1 km = 1000 m)

In Excel, start the formula with =
Use the mouse to click on the real distance (3.5)
Enter *1000
(to multiply by 1000)
Press Enter

Your spreadsheet should look like this:

Ride Number	Real Distance (km)	Scale Factor	Real Distance (m)	Real Distance (cm)	Distance on Map (cm)
1	3.5	40000	3500		

b) Real distance in cm

Now calculate the Real Distance in cm: how do you convert from metres to centimetres? Multiply the distance in m by 100!
On the same row, but under the 'Real Distance (cm)' label, enter a formula to multiply the real distance in m by 100

In Excel, start the formula with =
Use the mouse to click on the real distance in m (3500)
Enter *100
Press Enter

Step 3 - Find the map distance

Finally, calculate the Map Distance in centimetres: how do you convert from real distance to map distance? Divide the scale factor!
On the same row, but under the 'Distance on Map (cm)' label, enter a *formula* to divide the real distance by the scale factor.

In Excel, start the formula with =
Use the mouse to click on the real distance in cm (350000)
Enter /
Use the mouse to click on the scale factor (40000)
Press Enter

Your spreadsheet should look like this:

Ride Number	Real Distance (km)	Scale Factor	Real Distance (m)	Real Distance (cm)	Distance on Map (cm)
1	3.5	40000	3500	350000	8.75

Now repeat the calculations for the rest of the figures in the table.

Save your worksheet with the filename bike.
Print a copy of your worksheet.

6.3 Plans

Architectural drawings and plans of rooms use ratios so that they can be drawn accurately to scale. Many kitchen planners use computers to draw both the plans and representations of the appearance of the new cupboards and appliances within the kitchen.

Example 2

A rectangular kitchen is 2.0 m by 2.9 m, as shown below. Jenny is planning to buy some new kitchen floor cupboards to fit either side of her cooker along the wall opposite the door. She wants the cooker (which measures 60 cm by 60 cm) to be at least 50 cm away from the wall with the window. Standard sized kitchen cupboards are 60 cm by 60 cm, and any gap should be at the end of the room away from the window. Design a suitable layout to a scale of 1:50 and find the number of cupboards which she will need to buy.

Solution

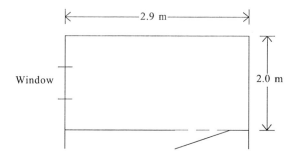

If the wall is 2.9 m long and the cooker is 60 cm long, then, using metres,

Length to be filled by cupboards = 2.9 - 0.6 m
= 2.3 m

Number of cupboards = $\dfrac{\text{Length to be filled}}{\text{Width of standard cupboard}}$

= $\dfrac{2.3}{0.6}$

= 3 remainder 0.5 m

So, she will need 3 cupboards 60 cm wide, and this will leave a gap of 0.5 m

Now draw the plan using a scale of 1:50

The kitchen measures 2.9 m long by 2.0 m wide

On the plan, the room length = $\dfrac{2.9}{50}$ m long

= $\dfrac{2.9}{50} \times 100$ cm long

= 5.8 cm long

On the plan, the room width = $\dfrac{2.0}{50} \times 100$ cm wide

= 4 cm wide

The cooker is 60 cm wide by 60 cm deep

On the plan the cooker width = $\dfrac{60}{50}$ cm wide

= 1.2 cm wide

Similarly, the cupboards will measure 1.2 cm wide on the plan.

The plan below shows a possible solution, with the cupboards arranged as shown:

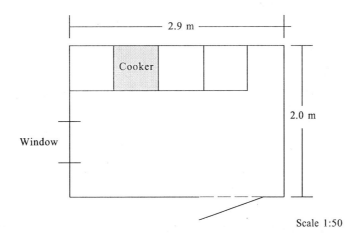

Scale 1:50

6.4 Three-dimentional Representation

Often it is useful to give a two-view representation of a three-dimensional object, as this helps to convey the depth of an object.

Example 2:

Draw a three-dimensional sketch of the cupboards and cooker in the kitchen above.

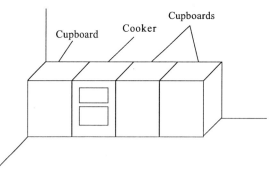

The fronts of the cupboards and cooker are to scale.

Exercise 6.2

1. Measure the floor area of a computer room, then measure the dimensions of the tables and cupboards in the room. Estimate the space needed for a person to sit at one of the tables. Create a plan of the current layout of the room and furniture, all to a suitable scale. Is there enough space to have more tables in the room?

2. Make a three-dimensional drawing of a 3½ inch floppy disc.

3. Make a three-dimensional drawing of a computer printer.

4. Measure the floor area of your kitchen, then measure the dimensions of all the cupboards, furniture and appliances in the room. Create a plan of the room, showing the position of its major contents, all drawn to a suitable scale.

Chapter 7
Percentages

At the end of this chapter you will be be able to

- [ ] convert a percentage
- [ ] express values as a percentage
- [ ] find a percentage of a quantity
- [ ] increase and decrease an amount by a given percentage

7.1 What is Meant by a Percentage?

A percentage is a form of numerical expression used to show a fractional part of a quantity which is assumed to have 100 units when it is complete, i.e. it is the number of parts out of 100,

e.g. 28% means 28 out of 100.

To convert a decimal or a fraction to a percentage, multiply it by 100.

Example 1: Convert 0.41 to a percentage.

Solution: To multiply the decimal quantity by 100 we have

$0.41 \times 100 = 41\%$

Example 2: Convert $\frac{23}{40}$ to a percentage.

Solution: To multiply the fraction by 100 we have

$$\frac{23}{40} = \frac{23}{40} \times 100$$
$$= \frac{23 \times 100}{40}$$
$$= 57.5\%$$

Exercise 7.1

Convert the following decimals and fractions to percentages:

1. 0.72   2. 0.375   3. 1.23   4. 0.084   5. 2.05   6. 0.729

7. $\frac{1}{2}$ 8. $\frac{3}{5}$ 9. $\frac{3}{20}$ 10. $\frac{7}{40}$ 11. $\frac{17}{50}$ 12. $\frac{7}{25}$

7.2 Expressing Values as a Percentage

Again, to convert a quantity to a percentage, multiply by 100.

Example 3:
Two students in a class of 16 are left-handed.
Express this as a percentage.

Solution: Expressed as a fraction of the whole class, the number of left-handed students is $\frac{2}{16}$

To express this fraction as a percentage, multiply by 100

i.e. percentage of left handed students $= \frac{2}{16} \times 100$

$= \frac{2 \times 100}{16}$

Percentage of left handed students $= 12.5\%$

Example 4:
11 students in a class of 16 have blue eyes.
Express this number as a percentage.

Solution: Expressed as a fraction of the whole class, the number of students with blue eyes is $\frac{11}{16}$

To express this fraction as a percentage, multiply by 100,

i.e. percentage of blue-eyed students $= \frac{11}{16} \times 100$

$= \frac{11 \times 100}{16}$

Percentage of blue-eyed students $= 68.75\%$

Exercise 7.2

1. Sarah gets 28 marks out of 80 in a test. Express this as a percentage.

2. Ben's pocket money goes up from £2 to £2.40.

    a) How much more does Ben receive each week?
    b) Express this increase as a percentage of his original pocket money.

3. Chris owns 14 CDs, and 4 of these are heavy metal. Find the percentage of CDs that are heavy metal.

4. Jon earns £205 and pays £36 of this as income tax. Express the income tax as a percentage of Jon's wage.

5. Jenny buys a bag of 9 oranges and gives 2 of them away. What percentage does she give away?

7.3 Finding a Percentage of a Quantity

To find the value of a percentage of a quantity, we express the percentage as a fraction or decimal first.

Example 5:
A class contains 25 students, and 60% are boys. How many boys are there in the class?

Solution: Express 60% as a part out of 100, i.e.

$$60\% = \frac{60}{100}$$

$$\text{The number of boys} = 25 \times \frac{60}{100}$$

$$= \frac{25 \times 60}{100}$$

$$= 15$$

i.e. 15 of the students are boys

Example 6:
You earn £45 and decide to save 25%. How much do you save?

Solution:

$$25\% = \frac{25}{100}$$

$$\text{Amount saved} = 45 \times \frac{25}{100}$$

$$= \frac{45 \times 25}{100}$$

$$= £11.25$$

i.e. you save £11.25

Exercise 7.3

1. Find:

   a) 23% of 50   b) 15% of 120   c) 45% of 20   d) 6.8% of 70

2. The total number of marks in a test is 60. If the pass mark is 40%, how many marks must a student get to pass?

3. 35% of students attending a college travel by bus. If there are 880 students in the college, find how many students travel by bus.

4. In a survey, 32% of students said their favourite drink was coke. If 125 students took part in the survey, how many students did this percentage represent?

5. Phil buys a consignment of fruit weighing 400 kg. If 13% of the fruit was bad, find the weight of bad fruit.

7.4 Increasing an Amount by a Given Percentage

An increase of 15% in an amount means that the amount is increased by $\frac{15}{100}$. The new amount will be equal to the original amount plus the 15% increase.

Example 7:
You earn £32 per week, and receive a 15% wage rise. What is your new wage?

Solution: The original amount of £32 represented 100. Your wage is being increased by 15%

$$\text{The increase} = 32 \times \frac{15}{100}$$
$$= \frac{32 \times 15}{100}$$
$$= £4.80$$

$$\text{The new wage} = \text{The original wage} + \text{Increase}$$
$$= 32 + 4.80$$
$$\text{New wage} = £36.80$$

Example 8:
The price of a computer printer is £186 plus VAT at 17.5%. What is the price of the printer including VAT?

Solution: The original printer price of £186 represents 100%. VAT of 17.5% is added to this price.

$$\text{The amount of VAT} = 186 \times \frac{17.5}{100}$$
$$= \frac{186 \times 17.5}{100}$$
$$= £32.55$$

$$\text{Price including VAT} = \text{Original price} + \text{VAT}$$
$$= 186 + 32.55$$
$$\text{Price including VAT} = £218.55$$

7.5 Decreasing an Amount by a Given Percentage

Sometimes the price of goods is decreased by a given percentage, e.g. during sale time when prices may be reduced, or if a discount is given. Again, the original price of the goods is the 100% figure. A decrease of 12% in a value means that new value will be equal to the original value minus this decrease.

Example 9: In a sale, a jacket which cost £42 is reduced by 12%. Find the sale price.

Solution: The original price of £42 represented 100%. The reduction is 12%, and this is taken off the original price.

$$\text{Reduction} = 42 \times \frac{12}{100}$$

$$= \frac{42 \times 12}{100}$$

$$= £5.04$$

Sale price = Original price - Reduction

$$= 42 - 5.04$$

Sale price = £36.96

Example 10:
A music shop offers a discount of 7.5% to all customers who belong to their Music Club. Find how much a Music Club member would pay for a CD which would normally cost £14.50.

Solution: The original price of £14.50 represented 100%. The discount is 7.5%, and this is taken off the original price.

$$\text{Discount} = 14.50 \times \frac{7.5}{100}$$

$$= \frac{14.50 \times 7.5}{100}$$

$$= £1.09$$

Discounted price = Original price - Discount

$$= 14.50 - 1.09$$

Discounted price = £13.41

Exercise 7.4 - Percentage Change

1. Lee was earning £23 for his Saturday job, and had a pay increase of 20%. How much did he earn now?

2. A restaurant adds a 10% service charge to all its meals. If the meal cost £9.20, how much do I pay?

3. A computer VDU cost £86. If VAT at 17.5% is added, how much would a customer have to pay?

4. The wage bill for a small business is £120,000. If the company awards all its employees an increase of 3.5%, what will the new wage bill for the company be?

5. During a sale, a clothes shop reduces the price of its shirts by 20%. If a shirt cost £12.80 originally, how much will it cost in the sale?

Exercise 7.4 (continued)

6. A computer shop offers a discount of 12% if a customer buys three or more computer games at a time. Melanie buys three games which cost £29.50, £35.50 and £44.75.

    a) What is the total cost of the games before any discount?
    b) How much will Melanie pay after the discount?

7. After going on a diet, a man's weight decreases by 18%. If he weighed 76 kg before, how much does he weigh after the diet?

8. George bought a motorbike for £2560. After a year, its value had decreased by 15%. How much is it worth now?

9. Harry buys his decorating materials from a company which offers 8% discount if he pays by cash. If the materials cost £28.45, how much will he pay?

10. The number of students enrolled on a new computer studies course goes up from 45 in the first year to 51 in the second year. Find the increase in the number of students enrolled as a percentage of the number in the first year.

Activity 6 - Spreadsheet Exercise
Increasing and Decreasing by a Percentage

You are asked to re-price a number of items in a shop, and have been given the item, the original price and the percentage change. You are to find both the amount of change and the new price. The items and figures you are asked to re-price are:

Item	Original Price (£)	Percentage Change (%)
CD	7.60	+20
Jacket	35.00	-10
Tape	8.60	-15
Shirt	14.50	+12.5
Jeans	31.60	+17
Shoes	42.60	-12.5
Socks	2.25	-30
Book	8.75	+10

This exercise allows you to increase or decrease an original value by a particular percentage. You will need to start with the original price and the percentage change.

Enter the data as follows:

Item	Original Price	Percentage Change (%)	Amount of Change	New Price
CD	7.60	20		

In Excel, start with =
Click on the original price for the item (7.60)
Enter *   *(to multiply)*
Click on the percentage change for the item (20)
Enter /100
(to divide by 100)
Press Enter

The amount of change = 7.60 × $\frac{20}{100}$, and so this is the formula we need to use.
Under Amount of Change, enter a formula to calculate this amount.

Now use a formula to find the total amount, by adding the original price and the amount of change.

In Excel,
Start the formula with =
Click on the original price for the item (7.60)
Enter +   *(to add)*
Click on the amount of change (1.52)
Press Enter.

Your spreadsheet should look like this:

Item	Original Price	Percentage Change (%)	Amount of Change	New Price
CD	7.60	20	1.52	9.12

Now repeat the calculations, using similar formulae, for all the other items. Save your worksheet using the filename Prices. Print a copy of your worksheet.

Exercise 7.5: Miscellaneous percentages

1. The value of a car decreases from £3420 to £2850 after one year.

    a) How much has the price decreased?
    b) Express this decrease as a percentage of the original amount.

2. In a school, 245 students enter an exam. 38 students gain a grade C.
   Express this number as a percentage of those entered.

3. If Greg's wage increases from £320 to £360 per week, find the percentage increase.

4. A salesman receives a bonus of 4% for the value of goods he sells above £2000. If he sells goods valued at £3560, what bonus will he receive?

Exercise 7.5 (continued)

5. The cost of making a toy was £7000 for 4000 toys. The manufacturer made a profit of 20% when she sold all the toys to a toy shop. The shop sold all the toys at £2.80 each.

   a) How much did it cost to make one toy?
   b) What price per toy did the manufacturer sell the toys for?
   c) How much profit did the manufacturer make altogether?
   d) How much profit did the shop make per toy?
   e) What percentage profit did the shop make?

6. A greengrocer buys 400 lb of strawberries for £180. He sells 200 lb at 70p per pound. He then reduces the price to 50p per pound, and sells 160 lb at this price. The rest of the strawberries go bad and he has to throw them away.

   a) Find the amount of money he obtained by selling the strawberries.
   b) How much was his profit?
   c) Express this profit as a percentage of what he paid.

7. A radio is advertised in a shop as having a list price of £70 plus VAT at 17.5%. The shopkeeper offers a discount of 18% before adding on the VAT. Calculate:

   a) The list price including VAT.
   b) The amount of discount before VAT is added.
   c) The final price of the radio, with VAT added, to the nearest penny.

8. In a sale, a computer desk which normally costs £69 was reduced by 8%, and a swivel chair which normally costs £78 was reduced by 12%. If Malcolm buys both a desk and a chair:

   a) How much did the desk cost after the reduction?
   b) How much did the swivel chair cost after the reduction?
   a) How much will he pay after the price reduction?
   b) What is the total price reduction on the desk and chair?
   c) Express this reduction as a percentage of the original cost of a desk and chair.

9. A college Drama Club produced a show, to be performed in the Assembly Hall. For each performance, the Hall had 300 seats costing £5 and 240 seats costing £4 each. For the first performance, 75% of the £5 seats were sold, and 60% of the £4 seats were sold.

   a) How many £5 seats were sold?
   b) How many £4 seats were sold?
   c) What was the percentage of all seats in the Hall sold?
   d) How much money was taken for the performance?
   e) How much money would be taken if all the seats were sold?
   f) Express the takings for the first night as a percentage of the takings if all the seats were sold.

10. An examining board had 8721 entries in 1994 and 9600 in the following year.

    a) Find the percentage increase in entries, correct to 1 decimal place.
    b) If the increase for the year 1996 was 9% of the entry in 1995, find how many entries there were in 1996.

Chapter 8
Presentation of Data

At the end of this chapter you will be be able to

- [ ] describe the types of variables
- [ ] draw and describe multiple bar charts
- [ ] draw and describe stacked bar charts
- [ ] draw and describe pie charts

8.1 Introduction

Statistics is about collecting and analysing data. Statistics are used widely by businesses, government, scientific institutions and other organisations.

Once relevant data has been collected, it is classified and analysed in order to make clear the important facts about the data. Often data is displayed graphically or as pictures, as the relevant facts can be interpreted more clearly this way.

8.2 Raw Data

Raw data is the original data collected, before it has been analysed or put into order.

The scores in a maths test were recorded as follows:

```
7    8    9    8    6    7    8   10    8    6
5    8    9    6   10    4    8    9    7    8    8    5    7
7    8    9    5    7
```

This data is in its original, raw, unordered form.

8.3 Tally Chart

Once data has been collected, it needs to be grouped; a tally chart records every occurrence of each value for the whole range of values.

First of all, scan the raw data to find both the lowest value and the highest value. This gives the range we must use. In the table above, the lowest value is 4 and the highest value is 10.

List these values in order down the page, as in the Tally chart below.

Go through each of the values in the raw data in turn, and put a tally mark (|) against the relevant value in the list. To help in totalling the scores quickly, the tally marks are usually arranged in groups of five. After the fourth tally mark (////) the fifth stroke is drawn through the previous four, i.e. ̶/̶/̶/̶/

Marks	Tally	Frequency
4	/	1
5	///	3
6	///	3
7	̶/̶/̶/̶/ /	6
8	̶/̶/̶/̶/ ////	9
9	////	4
10	//	2

Once all the tally marks have been recorded, fill the frequency column on the right.

We can see that one person scored 4 marks, three scored 5 marks, three scored 6 marks and so on. Nine people scored 8 marks, and this was the mark with the highest frequency. The total number of people taking the test was the total of all the frequencies, and this was 28.

8.4 Frequency Distribution

The frequencies found in the Tally chart above can be shown in a **frequency distribution**. This is displayed either as a bar chart or in a line graph. In the bar chart, the bars are of equal width, and the heights of the bars represent the actual frequencies. In a line graph, the frequences plotted are joined by a line.

In both cases, the number of Marks goes along the x-axis, and the Frequency goes along the y-axis. A title, which describes the data, should be included.

The frequencies from the Tally chart are plotted below in both a bar chart and line graph.

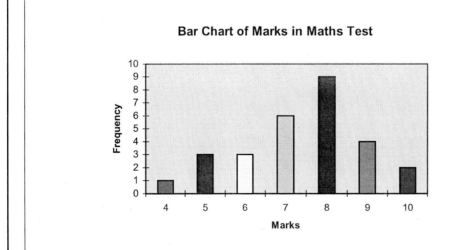

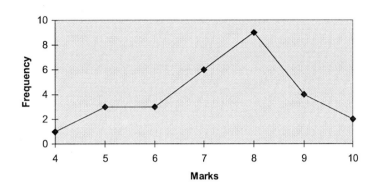

8.5 Types of Variables - Continuous and Discrete

Some variables such as height, weight and length can take a wide range of values. For example, the length of a piece of wood can be measured as 121 cm, or 120.8 cm or 120.7954 cm, depending on the accuracy of the measurement. Variables found by measuring are called **continuous** variables.

A **discrete** variable is one which can only take certain values. An example is the number of telephone calls a day received by a help desk, since the number must be a whole number. Another example would be clothing sizes - shoe size is a discrete variable, since it can only take certain values, e.g. 4, 4½, 5 etc. The various values of a **discrete** variable can usually be found by counting, whether it is the number of telephone calls received, or the number of people with a particular shoe size.

8.6 Grouped Distributions

All **continuous** data has to be grouped, e.g. if we were weighing people, we might group the weights in ranges such as 100 and less than 105 lb., 105 & < than 110 lb., 110 & < 115 lb. etc.

When the data is discrete, the data can be plotted against the individual, discrete value. However, if there are a lot of discrete values, it is better to group the data. In the example used previously, the range of marks was from 4 to 10, so there was only a small number of possible values of marks which a student could obtain. The marks in an end of year exam for example, could range up to 100, and so a grouped frequency distribution would be used. Normally, the range of data is divided into about six to twelve groups.

Spreadsheet Example 4
Plot the frequency distribution for the frequencies given in the Tally chart above as a Bar Chart

Enter the data into a new worksheet, to create a table of values for the Marks and Frequency, as follows:

	Marks	Frequency	
	4	1	
	5	3	
	6	3	
	7	6	
	8	9	
	9	4	
	10	2	

Now create a bar chart, with the Marks along the x-axis and the Frequency along the y-axis. Label your axes, and give the graph a Title.

In Excel, the ChartWizard takes you through 4 steps to create a graph.
Highlight the two columns of data, including the column headings ('Marks' and 'Frequency').
Now click on the Chart Wizard with the left mouse button.
Move the mouse back into the worksheet; use the left button to anchor the rectangular area where you want the graph to be. Now go through the 4 steps:

1. *Confirm or change the range of cells to be graphed - click **Next** button if correct.*
2. *Choose the chart type - click on **Column**, then on **Next** button*
3. *Choose the format of the chart type - click on **1** then on **Next** button*

4. ***Carefully check*** *that the preview of the graph is how you expect the final graph to be*
   a) *Rows or Columns - check that Columns is selected)*
   b) *Click on the **Next** button*

5. *Add Titles and axes names to your chart*
   *Under the **Legend** tab, de-select the **Show Legend** box - your column chart does not need one*
   *In the **Chart Title** box, enter the title **Frequency Distribution of Maths Marks***
   *In the **Axis Titles - Category (X)** box, enter **Marks***
   *In the **Axis Titles - Category (Y)** box, enter **Frequency***
   *Click on the **Finish** button, to display your graph.*

Spreadsheet Example 5
Plot the frequency distribution for the frequencies given in the Tally chart above as a Line Graph

Use the data from Spreadsheet Example 1. Create a Line Graph, with the Marks along the x-axis and the Frequency along the y-axis. Label your axes, and give the graph a suitable Title.

In Excel, follow the instructions given in Spreadsheet Exercise 1, but make the following modifications to steps 1, 2, and 3 of the Chart Wizard:

2. Choose the chart type - click on **Line**, then on **Next** button

3. Choose the format of the chart type - click on **1** then on **Next** button

5. Add Titles and axes names to your chart
   Under the **Legend** tab, de-select the **Show Legend** box - your column chart does not need one
   In the **Chart Title** box, enter the title **Line Graph of Maths Marks**
   In the **Axis Titles - Category (X)** box, enter **Marks**
   In the **Axis Titles - Category (Y)** box, enter **Frequency**
   Click on the **Finish** button, to display your graph.

Exercise 8.1

1. For each of the following, identify whether the variable is discrete or continuous:
   a) The number of PCs in a computer room
   b) The weight of a computer keyboard
   c) The temperature of the water in a swimming pool
   d) The number of sheets of paper used in an assignment
   e) The length of time a music track plays for
   f) The number of passengers in a car

2. A hospital maternity department records the birth weights of babies in kg. during a particular week as follows:

2.73	3.25	4.10	3.56	2.95	3.05	3.35	3.28
3.97	3.45	3.57	2.83	3.65	2.41	3.78	2.52
3.17	2.76	3.29	4.25	3.23	3.64	3.12	3.31
4.05	3.49	2.93	3.65	2.85	3.83	3.24	3.09
3.91	2.21	3.84	3.02	3.51			

    Create a tally chart of the weights, using ranges 2.2 & < 2.4, 2.4 & < 2.6 etc. Then plot a frequency distribution of your data as a line graph.

3. A company checked staff sickness records for the past year; the number of days sickness taken by each of its employees is given below:

0	2	1	2	7	0	3	0	2	1
0	5	2	9	1	2	1	9	2	3
1	0	2	2	3	1	4	8	0	2
1	4	1	5	4	3	0	2	1	0

   Form a tally chart of the number of days sickness. Then plot your data on a frequency distribution as a bar chart.

4. The times taken by children to run 200m on a school sports day were recorded in seconds as follows:

28.4	31.5	29.7	34.8	29.5	32.5	30.2	37.1
29.8	29.3	24.6	31.9	27.6	32.3	29.6	34.7
29.7	38.6	28.9	29.3	28.4	30.3	25.1	29.5
25.7	30.8	37.5	28.8	31.4	29.7	38.4	32.5

   Form a tally chart of the time taken, using the ranges 24 & < 26 seconds, 26 & < 28 seconds etc. Use the frequencies from your tally chart to plot a frequency distribution as a line graph.

5. Create a bar chart to show the number of mobile phone subscribers for the years 1992 - 1996, using the figures below:

Year	92	93	94	95	96
Subscribers (in thousands)	1190	1480	2320	4100	6100

6. More glass is being re-cycled nowadays. The table below shows the re-cycling rates expressed as a percentage, for several EU countries. Plot the data as a bar chart.

Country	Netherlands	Germany	Denmark	Italy	France	Greece	UK
% Re-cycling rate	80	75	60	50	47	35	25

8.7 Pie Charts

In a pie chart, the complete circle represents the total or whole of something. Each sector or slice represents that component's proportion of the whole. Pie charts are very useful in showing the proportions which each part takes, but it is not easy to find the total quantity which is being represented.

Example 1
In a particular class, 8 students travel to college by train, 10 by bus, 7 walk and 5 students cycle. Show this data in a pie chart.

Solution

First of all, calculate the angle which each sector has at the centre of the circle.
There are 360° at the centre of the circle.
Divide the 360° by the total students in the class, to give the angle for one student.
Then multiply this figure by the number of students using each mode of transport. We can also find the percentage of students using each mode of transport.

Total number of students = 8 + 10 + 7 + 5
= 30

Angle for 1 student = $\frac{360}{30}$ = 12°

The 8 students travelling by train will be represented by 8 × 12° = 96°, and so on for the other modes of transport.

Percentage of students travelling by train = $\frac{8}{30}$ × 100

= $\frac{8 \times 100}{30}$

= 26.67%

All the figures are shown in the table:

Mode of Transport	Number of Students	Angle at centre	Percentage travelling this way
Train	8	8 × 12° = 96°	$\frac{8}{30}$ × 100 = 26.7%
Bus	10	10 × 12° = 120°	$\frac{10}{30}$ × 100 = 33.3%
Walk	7	7 × 12° = 84°	$\frac{7}{30}$ × 100 = 23.3%
Cycle	5	5 × 12° = 60°	$\frac{5}{30}$ × 100 = 16.7%
Total	30	360°	100%

Now use this data for the Pie chart:

Mode of Transport Pie Chart

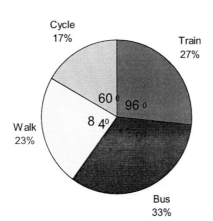

In this pie chart, both the angles at the centre *and* the percentage of students taking each mode of transport is shown.
It is easier to use **angles** in hand drawn pie charts.
Percentages are used in computer drawn pie charts.

Spreadsheet Example 6
Plot the Pie Chart using the Figures Given Above

Enter the data into a new worksheet, to create a table for the Mode of transport and Number of students, as follows:

	Mode of Transport	Number of Students	
	Train	8	
	Bus	10	
	Walk	7	
	Cycle	5	

Now create a pie chart, and give your chart a Title.

In Excel, the ChartWizard takes you through 4 steps to create a graph.
Highlight the two columns of data, including the column headings ('Mode of Transport' and 'Number of Students').
Now click on the Chart Wizard with the left mouse button.
Move the mouse back into the worksheet; use the left button to anchor the area where you want the graph to be. Now go through the 4 steps:

1. Confirm or change the range of cells to be graphed - click **Next** button if correct.
2. Choose the chart type - click on **Pie**, then on **Next** button
3. Choose the format of the chart type - then click on the **Next** button
4. **Carefully check** that the preview of the graph is how you expect the final graph to be
   a) Rows or Columns - check that Columns is selected
   b) Click on the **Next** button
5. Add a Title to your chart
   Under the **Legend** tab, de-select the **Show Legend** box - your pie chart does not need one
   In the **Chart Title** box, enter the title *Mode of Transport Pie Chart*
   Click on the **Finish** button, to display your graph.

Exercise 8.2

1. A shop recorded the sales of its PCs over the last 5 weeks as 8, 10, 9, 15 and 12. Show this data in a pie chart.

2. A travel agent recorded the number of holidays sold for various destinations, as shown below. Draw a pie chart and comment on the figures.

Destination	France	Italy	Spain	Greece	USA
No. of Holidays sold	56	84	105	120	21

Chapter 9
Averaging Data

At the end of this chapter you will be be able to

- [ ] calculate the arithmetic mean
- [ ] find the median
- [ ] find the mode
- [ ] compare the three averages

9.1 Introduction

When we do a survey and collect data we can represent the range and values graphically. Sometimes we might need to compare one group of data with another. To compare groups of data, we need a typical value from each group. The typical figure must be representative of the data from which it comes, and is an average. The most common averages are the arithmetic mean, the mode and the median.

9.2 Arithmetic Mean

The arithmetic mean is found by adding together all the data values, and then dividing this total by the number of values that make up the data, i.e.

$$\text{Mean} = \frac{\text{Total of all the data values}}{\text{Number of values}}$$

Example 1
The number of students attending a college badminton fun session on seven successive times was 12, 15, 9, 12, 11, 14, 10. Calculate the mean number attending.

Solution

Total of all the data values	=	$12 + 15 + 9 + 12 + 11 + 14 + 10$
	=	83
Number of values	=	7
Mean number attending	=	$\frac{83}{7}$
	=	11.86

This means that, if the same number of people had attended all seven sessions, then there would have been 11.86 students at each session.

Note: Whilst it is not possible for 11.86 people to attend, the decimal part is included to show that the mean is a calculated value.

9.3 Median

The median is the middle value when all the values are arranged in order. First the data must be arranged in order.

Example 2
Use the figures in Example 1 to calculate the median number of students attending a college badminton fun session. The numbers attending were 12, 15, 9, 12, 11, 14, 10.

Solution

Arranging the number of students in order gives

$$9, 10, 11, 12, 12, 14, 15$$

The middle value in the range is 12,

i.e. Median number of students = 12

9.4 Mode

The mode of a set of data is the data value which occurs most often. It is possible to have more than one mode in a set of data - in this case the data is said to be bimodal.

Example 3
Calculate the modal number attending the badminton fun session, using the figures from Example 2.

Solution

The number attending in order is 9, 10, 11, 12, 12, 14, 15

The attendance figure which occurs most often is 12,

i.e.   Modal attendance figure = 12

9.5 Comparisons and Benefits of Mean, Mode and Median

Each of the mean, mode and median are useful in giving an indication of the average of values in a set of data, but the particular benefits and relevance of each one are:

* The Mean is the average which is most widely used, and all the data values are used in order to calculate it. However, it is affected greatly by extreme values.

* The Mode is particularly useful with non-numeric data, e.g. type of transport to college, i.e. train/bus/bike/car/walk, or with shoe or shirt sizes.

* The median is very useful when there are extremes of values, as it is not affected by them.

9.6 Range

The range is a measure of the spread of the data. The range is larger when the data is more spread.

Example 4
Use the data from Example 1, where the number of students attending a college badminton fun session on seven successive times was 12, 15, 9, 12, 11, 14, 10. Calculate the range of the number of students attending.

Solution

From the data, the largest number of students attending = 15
the smallest number of students attending = 9
Range = 15-9
i.e. Range = 6 students

Note: The range is a very basic measure of spread of the data, because it only uses two values - the largest and the smallest - and ignores all the values in between.

Exercise 9.1

1. During an athletics session, the distances jumped by nine students in the long jump event were 4.80m, 4.05m, 4.20m, 4.30m, 4.45m, 4.20m, 4.35m, 4.10m, 4.25m to the nearest 0.05 m. Find:
    a) The mean distance jumped
    b) The modal distance jumped
    c) The median distance jumped
    d) The range of distance jumped

2. Keith records the amount of money he spends in the college refectory during one week as £2.34, £1.45, £1.53, £2.08, £1.80. What is the mean amount he spends per day?

3. The number of shirts sold during a week were recorded as follows:

Shirt size	$14\frac{1}{2}$	15	$15\frac{1}{2}$	16	$16\frac{1}{2}$	17	$17\frac{1}{2}$
Number sold	3	6	7	5	2	1	1

Find the modal size of shirt.

4. The price of word processors in a store were recorded as £223, £279, £339, £479, £259, £329, £199. Calculate the mean price and the range of prices recorded.

5. Some batteries were tested to find how long they would work, and the times recorded in hours were: 218, 220, 148, 235, 223, 225, 215. Find:
    a) The mean hours they worked
    b) The median hours they worked
    c) The modal hours they worked
    d) Which of these averages gives the best estimate of the hours life of a battery.

Exercise 9.1 (continued)

6. The number of marks gained in an exam were 58, 63, 75, 82, 47, 69, 62, 88. Calculate the mean exam mark and the range of marks gained.

7. The price of copiers on sale in one store were recorded as £253, £459, £869, £799, £229, £279, £675. Find:
    a) The mean price of copiers
    b) The median price of copiers
    c) The range of copier prices

8. The weights of new-born babies in kg. recorded one day were 3.3, 3.5, 2.8, 2.9, 4.0, 3.7, 3.6, 3.9, 3.4. Find the median weight.

9. A technician recorded the length of time in minutes taken to diagnose and mend faulty PCs as 25, 56, 125, 54, 78, 83, 24, 37, 74, 89, 92. Find:
    a) The mean time taken
    b) The median time taken

10. The price of PCs with pentium processors was quoted at various stores as £999, £1099, £1249, £1179, £1320 and £1210. Calculate the mean price.

Chapter 10
Perimeters and Areas

At the end of this chapter you will be be able to find the perimeter and area of a:

☐ quadrilateral
☐ parallelogram
☐ rectangle
☐ triangle
☐ circle

10.1   Introduction

A **polygon** is a general name for any plane figure with straight sides.

A **triangle** is a plane figure which has 3 sides.

A **quadrilateral** is a plane figure which has 4 sides.

Parallelograms, **rectangles** and **trapezium** are all quadrilaterals with additional properties.

The **perimeter** is the length round the outside of an object, so add the lengths of all of the sides.

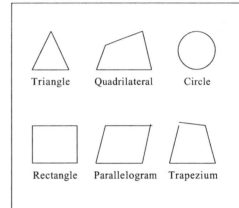

10.2   Perimeter of a Quadrilateral

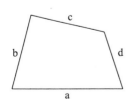

A quadrilateral is any plane figure with four sides.

Perimeter = distance around outside
= a + b + c + d

65

10.3 Perimeter of a Parallelogram

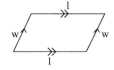

A parallelogram is a quadrilateral in which both pairs of opposite sides are parallel.

$$\text{Perimeter} = 2l + 2w$$

Rectangle

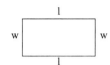

A rectangle is a parallelogram in which all the angles are 90°

$$\text{Perimeter} = 2l + 2w$$

Square

A square is a rectangle with all sides of equal length

$$\text{Perimeter} = 4w$$

Example 1
Calculate the perimeter of the flower bed shown below:

Solution

Length EF	=	AB + CD
	=	9 + 10
	=	19 ft
Length AF	=	BC + DE
	=	10 + 4
	=	14 ft
Perimeter	=	AB+BC+CD+DE+EF+FA
	=	9 + 10 + 10 + 4 + 19 + 14
Perimeter	=	66 ft

Exercise 10.1

1. Calculate the perimeter of the following:

   a) A rectangle with sides of length 12 cm by 5 cm
   b) A parallelogram with parallel sides of length 14 in and 8 in
   c) A square with sides of length 9 ft
   d) A quadrilateral with sides of length 8 m, 3m, 7m and 14m

2. Calculate the perimeter of the following shapes, where all dimensions are in cm:

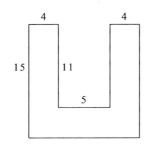 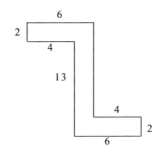 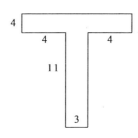

3. Calculate the perimeter of a rectangular jigsaw puzzle which has sides of length 38 cm by 24 cm.

4. A plan of Andrea's bedroom is shown opposite. She is redecorating the room, and plans to fix a decorative frieze at the top of the wall at ceiling height. Calculate the length of frieze she will need, if all measurements are in cm.

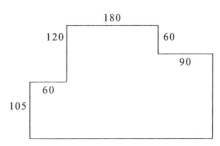

5. Brian is helping his father to lay a carpet. They will need to put gripper rod all round the edge of the room, to which the carpet is to be fixed, to keep its shape. What length of gripper rod should be bought, if they are carpeting the room shown opposite, and all dimentions are in feet?

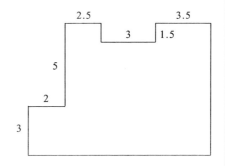

6. Damien plans to frame a photograph measuring 6 in by 8 in, with a card surround, which is 2 in wider all round than the photograph. He will put a frame round the outside of the card. Find the perimeter of:
a) The card surround
b) The photograph

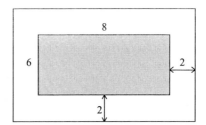

7. Carly is making a rectangular cushion for her grandmother, and plans to stitch a fancy binding round the edge of the cushion. If the cushion measure 45 cm by 37 cm, what length of binding will she need?

67

10.4 Area of a Rectangle

The area of a rectangle is its length multiplied by its width. The units used must be the same for both measurements.

Example 2
Calculate the area of the rectangle which is 8.2 cm long by 6.7 cm wide.

Solution

Area = length × width
 = 8.2 × 6.7 cm²
 = 54.94 cm²

To calculate the area of some composite shapes, split them up into rectangles and find the area of each rectangle. Sometimes, it is easier to find the area of one shape and then subtract the area of another.

Example 3
Find the area of the shape to the right, where all dimensions are in cm.:

Solution

This shape can be split into three rectangles, as shown below.

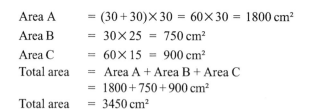

Area A = (30 + 30) × 30 = 60 × 30 = 1800 cm²
Area B = 30 × 25 = 750 cm²
Area C = 60 × 15 = 900 cm²
Total area = Area A + Area B + Area C
 = 1800 + 750 + 900 cm²
Total area = 3450 cm²

Alternatively, sometimes it may be easier to subtract one area from another. Here, the shape is split into a large rectangle Z and a small rectangle Y.

The area required = Area Z - Area Y
Area Z = (30 + 25 + 15) × (30 + 30)
 = 70 × 60 cm²
 = 4200 cm²
Area Y = 30 × 25 = 750 cm²
The area required = Area Z - Area Y
 = 4200 - 750
 = 3450 cm²

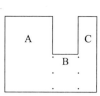

Example 4
A picture measuring 25 cm by 18 cm is mounted on a coloured border, which is 32 cm by 24 cm. Calculate the area of the border which is visible.

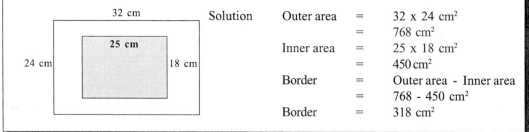

Solution
Outer area = 32 × 24 cm²
 = 768 cm²
Inner area = 25 × 18 cm²
 = 450 cm²
Border = Outer area - Inner area
 = 768 - 450 cm²
Border = 318 cm²

Exercise 10.2

1. Calculate the areas of the following rectangles:

    a) 12.3 cm x 8.4 cm
    b) 7.2 ft x 3.8 ft
    c) 5.9 in x 23.5 in
    d) 9.1 m x 4.8 m
    e) 32.1 m x 43.6 m
    f) 16.3 cm x 17.6 cm

2. The perimeter of a square is 24 cm. Calculate:

    a) The length of each side.
    b) The area of the square.

3. In a kitchen, a worktop measuring 60 cm by 220 cm has a hole measuring 46 cm by 90 cm cut into it to take a new sink. What is the area of worktop remaining?

4. Calculate the area of the shapes in Exercises 14.1 question 2.

5. Using the photograph and card surround in Exercises 14.1 question 6, calculate:

    a) The area of the photograph
    b) The area of the card surround

6. The area of a rectangular room is 17.76 m². If the length of the room is 4.8 m, find:

    a) The width of the room
    b) The perimeter of the room

7. Calculate the floor area of the room in Exercises 14.1 question 5.

8. Calculate the areas of the following shapes:

    a) All lengths in cm
    b) All lengths in inches
    c) All lengths in ft.

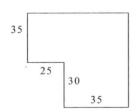
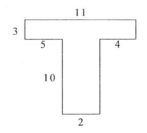
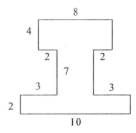

9. A house has a rectangular garden. The owner decides to landscape the garden so that it has an area for a patio, a lawn and a flower bed, as shown opposite. If all dimensions are in metres, calculate the area of:
   a) The patio
   b) The lawn
   c) The flower bed
   d) Check your answers by finding the total area of the garden

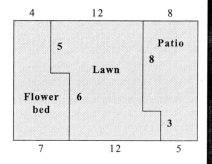

10. A 140 cm length of picture frame material is cut into four pieces to form the outside rectangular frame for a picture. If the length of one of the sides is 41 cm, calculate:

   a) The width of the frame
   b) The area enclosed within the frame

10.5 Area of a Parallelogram

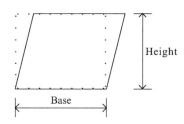

A parallelogram is like a rectangle which has been slanted to one side.

The rectangle equivalent to the parallelogram is shown by dotted lines.

Area of parallelogram = Base length × Perpendicular height

Note that the formula for the area of a parallelogram always uses the perpendicular (or vertical) height.

Example 5 Find the area of the parallelogram with a base length of 10 ft and height of 7 ft.

Solution

Area = base × height
    = 10 × 7 ft^2
    = 70 ft^2

Exercise 10.3

1. Find the areas of the following parallelograms:

    a) Base length 7.2 cm, height 5.1 cm.
    b) Base length 12.3 cm, height 4.8 cm.
    c) Base length 18.2 m, height 11.6 m.
    d) Base length 8.4 in., height 6.5 in.

2. The area of a parallelogram is 69 cm². If its base length is 12 cm, calculate the height.

3. The area of a parallelogram is 126.38 ft². If it has a height of 14.2 ft, calculate the base length.

4. Calculate the area of the shapes shown below, where all dimensions are in cm:

    a)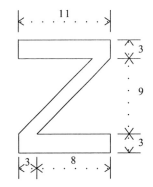
    b)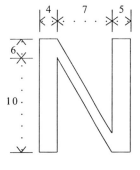

10.6 Area of a Triangle

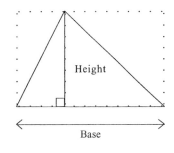

The area of a triangle can be found by enclosing it inside a rectangle.

The area of the triangle is half the area of the rectangle.

$$\text{Area of triangle} = \frac{1}{2} \times (\text{area of rectangle})$$

$$= \frac{1}{2} \times (\text{base} \times \text{height})$$

Example 6
Calculate the area of the triangle with base 8.2 cm and height 6 cm.

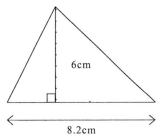

Solution

$$\text{Area of triangle} = \frac{1}{2} \times (\text{base} \times \text{height})$$
$$= \frac{1}{2} \times 8.2 \times 6 \text{ cm}^2$$
$$= 24.6 \text{ cm}^2$$

Exercise 10.4

1. Calculate the area of the following triangles:

    a) Base 11.5 cm and height 7.6 cm.
    b) Base 30 m and height 8.3 m.
    c) Base 12.8 ft and height 5.4 ft.
    d) Base 9.4 cm and height 3.9 cm.
    e) Base 18.3 m and height 10.2 m.

2. Calculate the height of the following triangles:
    a) Area 47.32 cm², base 10.4 cm
    b) Area 284.05 m², base 24.7 m
    c) Area 53.675 cm², base 11.3 cm

3. Calculate the area of the following shapes, where all the dimensions are in ft:

   a)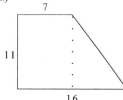

   b)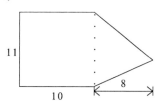

10.7 Circumference of a Circle

The diagram opposite shows the main parts of a circle. In a circle, the perimeter is always called the **circumference**.

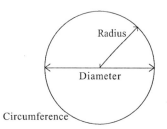

The value of

$$\frac{\text{Circumference}}{\text{Diameter}} = 3.14159...$$

This figure is given the symbol Π, (pronounced pi), and it is the Greek letter for p. The value of Π cannot be worked out exactly, but can be taken to be 3.142 to 3 d.p., or we can use the approximation $\frac{22}{7}$

Thus we can say that $\frac{\text{Circumference}}{\text{Diameter}} = \Pi$

and rearranging this formula,

Circumference = $\Pi \times$ Diameter

Since   Diameter = $2 \times$ Radius

then   Circumference = $2 \times \Pi \times$ Radius
i.e.   C = $2\Pi r$
or   C = Πd

where   C = circumference, r = radius and d = diameter

Example 7
Use $\Pi = \frac{22}{7}$ to calculate the circumference of a circle with diameter 14 cm.

Solution

$$C = \Pi d$$
$$= \frac{22}{7} \times 14 \text{ cm}$$
$$= \frac{22 \times 14}{7} \text{ cm}$$
$$= 22 \times 2 \text{ cm}$$
$$= 44 \text{ cm}$$

Example 8
Use $\Pi = 3.142$ to calculate the radius of a circle which has a circumference of 37 m.

Solution

$$C = 2\Pi r$$

i.e.
$$r = \frac{C}{2\Pi}$$
$$= \frac{37}{2 \times 3.142}$$
$$= 5.89 \text{ m to 2 d.p.}$$

Exercise 10.5

1. Using $\Pi = 3.142$ (or the Π button on your calculator), calculate the circumference of a circle which has a:

   a) Radius of 8.2 cm
   b) Radius of 12.4 m
   c) Diameter of 9.7 ft
   d) Diameter of 35.6 m
   e) Radius of 41.5 m
   f) Diameter of 23 cm
   g) Diameter of 18.1 cm
   h) Radius of 7.2 in

2. Use $\Pi = \frac{22}{7}$ to calculate the circumference of a circle which has a:

   a) Radius of 21 ft b) Radius of 0.7 m c) Diameter of 28 ft
   d) Diameter of 350 cm e) Radius of 14 in f) Radius of 42 cm

3. Calculate the radius of a circle which has the circumference measurement given below:

   a) 27 cm
   c) 18.6 ft
   b) 9.4 m
   d) 12.9 in

4. Calculate the diameter of a circle which has the circumference measurement given below:

   a) 19.7 in
   c) 8.5 m
   b) 13.6 cm
   d) 56.3 cm

5. A circular kitchen table has a radius of 62.5 cm. Calculate the circumference of the table.

6. The cover which protects a Winchester disk drive has a circumference of 116 cm. Calculate the radius of the cover.

7. Most floppy disks used nowadays are known as 3½" disks, meaning that their diameter is 3½". Calculate the circumference of the disks.

8. A circular running track is 330 m. Calculate the radius of the track.

9. The wheels on Bruce's bike have a diameter of 66 cm. Find:

   a) The distance round the outside of the wheel
   b) The number of revolutions that the wheel makes when Bruce cycles 1 km.

10. Geoff's model railway has a circular track which is mounted on a wooden board 4ft. square. If the outside edge of the track just touches all four sides of the wooden board, find:

    a) The radius of the outside edge of the track
    b) The circumference round the outside of the track.

10.8 Area of a Circle

It can be shown that the area of a circle is given by

Area of circle = Π × Radius²

Or A = Πr²

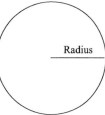

Example 9

Use $\Pi = \frac{22}{7}$ to calculate the area of a circle with radius of 21 cm.

Solution

A = Πr²

= $\frac{22}{7}$ x 21 x 21 cm²

= 22 x 3 x 21 cm²   *(after dividing by 7)*

= 1386 cm²

Example 10 - Use Π = 3.142 to calculate the radius of a circle which has an area of 27.4 in²

A = Πr²

27.4 = 3.142 x r²

r² = $\frac{27.4}{3.142}$

= 8.72

r = $\sqrt{8.72}$

r = 2.95 cm to 2 d.p.

Exercise 10.6

1. Use $\Pi = \frac{22}{7}$ to calculate the area of the circles which have radius of:

   a) 7 cm   b) 21 ft   c) 42 mm

2. Use Π = 3.142 to calculate the area of the circles which have radius of:

   a) 2.47 cm   b) 18.7 ft   c) 6.7 m

   d) 12.5 in   e) 23.1 cm   f) 7.2 cm

3. Use Π = 3.142 to calculate the radius of the circle which has an area of:

   a) 29.7 cm²   b) 35.6 ft²   c) 54.8 cm²

   d) 8.9 m²   e) 43.5 in²   f) 17.2 m²

Exercise 10.6 (continued)

4. A circular fish pond has an area of 27.4 m². Calculate

   a) The radius of the pond.
   b) The circumference of the pond.

5. A set of dinner mats comprises four serving mats each of diameter 30 cm and six place mats each of diameter 20 cm. Calculate:

   a) The area of a serving mat
   b) The area of a place mat
   c) The total area which the set of mats will cover.

Chapter 11
Volumes

At the end of this chapter you will be be able to find the volume of a:

- cuboid
- cube
- cylinder

11.1 Introduction

Volume is the size or amount of space which a solid figure takes up. It has three dimensions, and these are length, width and height.

Many shapes are prisms, i.e. they have a constant cross section. Typical prisms include cuboids, cubes and cylinders.

11.2 Cuboid

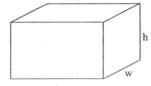

A cuboid is a solid which has six rectangular faces.

Volume = length x width x height

or

V = l x w x h

Note that all the measurements must be in the same units.

11.3 Cube

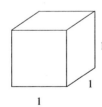

A cube is a cuboid in which all of its sides are the same length.

Volume = length x width x height

so

Volume = l x l x l

Example 1

Calculate the volume of a cuboid with length 18 cm, width 9.2 cm and height 3.4 cm.

Solution

Volume = length × width × height
= $18 \times 9.2 \times 3.4$ cm³
Volume = 563.04 cm³

Exercise 11.1

1. Calculate the volumes for the following cuboids, changing the units before calculation if necessary:

   a) Length 16 cm,   width 9 cm,   height 12.5 cm
   b) Length 24 m,   width 11 m,   height 6.4 m
   c) Length 117 mm,   width 3.4 cm,   height 7.6 cm
   d) Length 18 cm,   width 8.8 mm,   height 2.7 cm
   e) Length 1.8 ft,   width 7.6 ft,   height 3.4 ft
   f) Length 6.5 cm,   width 59 mm,   height 41 cm
   g) Length 5.2 m,   width 421 cm,   height 1.6 m

2. Calculate the volume of a cube which has sides of length 4.7 cm.

3. A cake tin is 23 cm long by 11.8 cm wide by 7.2 cm deep. Find the volume of cake that would fill the tin.

4. A swimming pool is 18 metres long by 8.5 metres wide. If the water is at a uniform depth of 1.4 metres, what is the volume of water in the pool?

5. Calculate the volume of a computer tower which is 17 in deep by 9 in wide by 17 in high.

6. Calculate the volume of a cube which has sides of length 13.4 in.

7. When Peter bought a visual display unit, it was surrounded by polystyrene and packed in a box measuring 48 cm long by 44 cm wide by 42 cm high. The volume of the vdu was 42735 cm³. Calculate:

   a) The volume of the box
   b) The volume of the polystyrene

8. A child's sand pit measures 2.2 m by 1.6 m. If the sand has a uniform depth of 23 cm, calculate the volume of sand.   (Take care with units).

11.4 Cylinders

A cylinder is a prism which has a circular cross section.

Volume of a cylinder = area of circle × height

Volume = $\Pi r^2 \times h$
= $\Pi r^2 h$

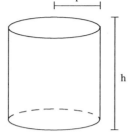

Example 2
Calculate the volume of a cylindrical water tank which has a radius of 6.7 cm and height of 35 cm, in both cm³ and litres. (1 litre = 1000cm³)

Solution

Volume = $\Pi r^2 h$
Volume = $3.142 \times 6.7 \times 6.7 \times 35$ cm³
= 4,936.55 cm³
Volume = 4,937 cm³
to the nearest cubic cm.

= $\frac{4937}{1000}$ litres

= 4.94 litres to 2 d.p.

Example 3
Calculate the radius of a cylinder if the volume is 736 in³ and the height is 5.2 in.

Solution

Volume = $\Pi r^2 h$
736 = $3.142 \times r^2 \times 5.2$
= $16.34 \times r^2$

r^2 = $\frac{736}{16.34}$

r^2 = 45.04

r = $\sqrt{45.04}$

r = 6.71 in to 2 d.p.

Exercise 11.2

1. Calculate the volume of the following cylinders:

    a) Radius 5 cm, height 16 cm.
    b) Radius 23mm, height 37 mm.
    c) Radius 46 cm, height 28 cm.
    d) Radius 32 cm, height 51 cm.

2. A cylindrical paint tin has a diameter of 15 cm. If the depth of the paint in the tin is 11 cm, find the volume of paint in the tin.

3. A tin of soup has a diameter of 7.5cm. If the height of the tin is 10.5 cm, find the volume of soup in the tin.

4. A cylindrical water tank can store 120 litres of water. If the height of the cylinder is 89 cm, find the radius of the tank. (1 litre = 1000 cm³)

5. A child's circular sand pit has a radius of 95 cm. If the sand is 25 cm deep, what volume of sand is there in the sand pit in m³?

6. A cylindrical water carrier has a radius of 8.5 cm and is 23 cm high. Find:

    a) The volume of water it will hold in cm³
    b) The volume of water in litres.

7. A full cylindrical paint tin has a diameter of 15 cm and height of 16 cm. Find:

    a) The volume of paint in the tin
    b) If the paint is poured into a tin with diameter of 17 cm, find the height which the paint would reach in the second tin.

Chapter 12
Assignments

Objectives

At the end of this chapter you will be able to practise the skills acquired using this workbook by carrying out a set of assignments

12.1 Assignment 1

Part A

Many students work part-time, either by delivering newspapers, working at a supermarket or baby-sitting. You are to design the data collection procedures for a survey to find out the wages earned by students from their part-time employment. You should collect details of hours worked and rate per hour, as well as the type of work and employer (i.e. supermarket or pharmacy or corner shop etc.) You are required to:

a) Explain why the data is being collected.

b) Write down the methods you will use to collect the data.

c) Write down how you plan to record the data and the questions which you will ask.

d) Design a questionnaire which you will use to record the data.

e) Conduct a survey of 15 - 20 students who work part-time and record their details accurately.

Part B

Calculate the average hours worked and the average wage earned part-time by the students on your survey, present the results graphically and state your conclusions. From the data you have collected:

a) Calculate the mean, median and modal number of hours worked each week by students employed part-time.

b) Calculate the mean, median and modal wage earned each week by students employed part-time.

c) From the range of values, use a tally chart to organise the data into about 5 to 8 wage groups.

d) Draw a frequency distribution of the figures found in (c).

e) Write down the number of students in the lowest wage group; express this number as a ratio of the total students in the survey. Then express the number in the lowest wage group as a percentage of the total students.

f) Write down the number of students in the highest wage group; express this number as a ratio of the total students in the survey. Then express the number in the highest wage group as a percentage of the total students.

(g) Display the numbers of students in each group (which you found in (c)) on a pie chart.

(h) Comment on the results you have found; what conclusions can be drawn from your figures?

12.2 Assignment 2

A manufacturer is planning to bring out a new range of fashion clothing and footwear, which will be sold in both this country and abroad. In order to plan for the sizing of these garments, you are asked to undertake a survey on the height and shoe size of both male and female students.

a) Determine the questions you will need to ask; then prepare a suitable data collection form, so that you can identify separately male and female data

b) Conduct a survey of at least 15 male and 15 female students

c) Calculate the mean, median and modal height of both male and female students

d) Convert the mean, median and modal height to imperial units, if the original measurements were metric. (Convert the measurements to metric units if you used imperial units for your original measurements)

e) Calculate the mean, median and modal shoe size for both male and female students.

f) Continental shoe sizes are different from those used in the UK. Use the table below to convert the shoe sizes found in (e) to their continental equivalent.

UK size	3	4	5	6	7	8	9	10
Continental size	36	37	38	39	40	41	42	43

g) Display the height data you have collected in a variety of forms (multiple bar chart, stacked bar chart, pie chart) for both male and female students

h) Display the shoe size data you have collected in a variety of forms

i) Write down your conclusions about male and female students' height

j) Write down your conclusions about male and female students' shoe size.

12.3 Assignment 3

Computer games are very popular nowadays, particularly among young people. A shop selling computer games has asked you to conduct a survey among the students at your college to find out both the three most popular computer games and the number of games owned by each person.

a) Prepare a data collection form which will be suitable for the above survey.

b) Conduct a survey of about 20 students.

c) Calculate the mean, median and modal number of games owned.

d) Use a tally chart to organise the data on the number of games owned into suitable groups; plot this data.

e) Determine the three most popular games.

f) Write down all the conclusions that can be drawn from your survey..

12.4 Assignment 4

Your college is planning to redecorate some of its rooms and needs to check the area of walls to be decorated in order to find out both the amount of paint needed and how many rooms can be decorated within this year's budget. To assist in this work, ensure that you:

a) Prepare a suitable data collection form for the above task.

b) Survey four rooms and measure the size of all the walls in them.

c) Calculate the area of wall to be decorated in each room.

d) Convert the dimensions of each wall to imperial units, if the original measurements were metric (or convert the dimensions to metric units if the original measurements were imperial).

e) If 1 litre of paint will cover approximately 12 sq.m or 130 sq.ft, use the answer you found in (c) to estimate the number of litres of paint needed.

12.5 Assignment 5

In preparation for an Open Day, a further education college has prepared some photographs to show students at work in the college's I.T department. The photographs will be mounted on card which will be displayed on hessian screens.

The photographs all measure 30 cm high × 40 cm wide, and are to be mounted on cardboard which will give a border 6 cm wide all round.

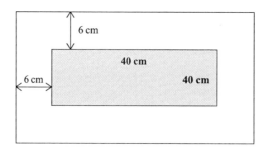

a) Calculate the area of each photograph

b) Calculate the area of cardboard on which one photograph will be mounted.

c) Find the area of the cardboard surround that will be visible.

d) Write down the ratio of the area of the photograph to the area of the card found in (b), and express this ratio in its simplest terms.

e) If each hessian screen is 1.6 m high by 1.5 m wide, calculate the area of the screen.

Six photographs, mounted on card, are to be displayed on each screen.

f) Draw a layout of how the photographs and card can be mounted on the screen.

g) What area of the screen will still be visible?

12.6 Assignment 6

A company is considering upgrading their PCs by replacing the cpu and memory. The prices quoted for cpu upgrades are £120, £180, £350 and £430; the prices quoted for memory upgrades are £60, £75, £125 and £135. Draw up a possibility space to show all the combinations. From this determine the probability that the total cost will be:

a) Less than £200.

b) More than £400.

c) Between £250 and £400.

Appendix | Answers

ANSWERS TO EXERCISES

Chapter 2 - Directed Numbers

Exercise 2.1

1. 12	2. -7	3. -7	4. -8	5. 4	6. -6	7. -11	8. 8
9. -1	10. -10	11. 5	12. -7	13. -1	14. 5	15. -2	

Exercise 2.2

1. -9	2. -10	3. 8	4. -7	5. 24	6. 5	7. 10	8. -8
9. -14	10. 5	11. 30	12. -10	13. 5	14. -20	15. 12	

Exercise 2.3

1. -18	2. 11	3. 21	4. -8	5. 6	6. 3	7. -10	8. -5
9. 12	10. 4	11. 14	12. 17	13. 5	14. 5	15. -3	16. 6
17. 8	18. 6						

Chapter 3 - Conversion of Units

Exercise 3.1

1. a) 0.396 kg   b) 1.78 lb   c) 0.54 kg   d) 3.78 lb
2. a) 14.56 litres   b) £9.16
3. a) 10.35 kg   b) 4.65 kg   c) 10.3 lb
4. 195 cm by 150 cm
5. 14.29 miles
6. 112.63 km/h
7. 1.8 m by 2.85 m
8. a) 75.62 km/gallon   b) 16.62 km per litre

Activity 4 - Spreadsheet Exercise

1. a) 37.78°C,   b) 48.89°C,   c) -17.78°C,   d) -6.67°C,   e) 10°C
2. a) 284°F,   b) 356°F,   c) 14°F,   d) 37.4°F,   e) 104°F

Exercise 3.2

1. a) 154.4 Ff   b) $108.5   c) 29925 Pesetas   d) 1,194,000 Lira   e) 130.35 Dm
2. a) £64.77   b) £209.38   c) £4.51   d) £16.13   e) £15.49
3. a) 15960 Pesetas   b) 4590 Pesetas   c) £23.01
4. £1.36
5. a) 463.2 Ff   b) 201.2 Ff   c) £26.06

Chapter 4 - Fractions

Exercise 4.1

1. a) $\frac{2}{4}$ b) $\frac{5}{8}$ c) $\frac{7}{20}$ d) $\frac{12}{32}$ e) $\frac{52}{72}$ f) $\frac{5}{9}$
2. a) $\frac{8}{12}$ b) $\frac{9}{21}$ c) $\frac{15}{40}$ d) $\frac{16}{72}$ e) $\frac{48}{60}$ f) $\frac{40}{48}$
3. a) $\frac{1}{2}$ b) $\frac{4}{5}$ c) $\frac{1}{3}$ d) $\frac{3}{5}$ e) $\frac{5}{6}$ f) $\frac{2}{3}$ g) $\frac{2}{3}$ h) $\frac{17}{20}$
4. a) $1\frac{1}{3}$ b) $1\frac{5}{8}$ c) $4\frac{1}{4}$ d) $2\frac{4}{5}$ e) $3\frac{2}{7}$ f) $4\frac{1}{3}$ g) $3\frac{7}{10}$ h) $4\frac{1}{2}$
5. a) $\frac{5}{3}$ b) $\frac{15}{4}$ c) $\frac{37}{7}$ d) $\frac{143}{20}$ e) $\frac{25}{9}$ f) $\frac{9}{2}$ g) $\frac{14}{5}$ h) $\frac{23}{6}$
6. a) $\frac{3}{8}$ b) $\frac{3}{16}$
7. $\frac{1}{8}$
8. $\frac{2}{9}$
9. a) $\frac{1}{4}$ b) $\frac{1}{7}$ c) $\frac{3}{28}$ d) $\frac{1}{2}$
10. a) 800 g b) $\frac{1}{4}$ c) $\frac{3}{8}$ d) $\frac{5}{16}$ e) $\frac{1}{16}$

Chapter 5 - Ratio

Exercise 5.1

1. a) 1 : 6 b) 1 : 5 c) 2 : 5 d) 2 : 3 e) 10 : 13
2. a) 2 : 7 b) 3 : 7 c) 1 : 5 d) 1 : 7 e) 1 : 5
3. 2 : 3
4. 4 : 3
5. 5 : 2
6. a) 4 : 6 b) 2 : 1 c) 20 : 25 d) 3 : 5 e) 7 : 12 f) 7 : 9
7. 3 : 4 : 5
8. 25 : 2

Exercise 5.2

1. £450, £270
2. 24, 36, 60
3. £90, £60, £150, £150
4. £18
5. £4.83, £2.07
6. £8
7. £17.50, £24.50
8. £45
9. 160 g, 280 g, 80 g
10. £45
11. 56 g
12. a) 96 b) 40

Chapter 6 - Maps, Scales and Plans

Exercise 6.1

1. 4.17 km
2. 0.68 km
3. 65 cm
4. a) 1 : 200,000 b) 12.4 km c) 9.6 km
5. 24 cm
6. 28.75 cm

Chapter 7 - Percentages

Exercise 7.1

1. 72%　　2. 37.5%　　3. 123%　　4. 8.4%　　5. 205%　　6. 72.9%
7. 50%　　8. 60%　　9. 15%　　10. 17.5%　　11. 34%　　12. 28%

Exercise 7.2

1. 35%
2. a) 40p　　b) 20%
3. 28.6%
4. 17.6%
5. 22.2%

Exercise 7.3

1. a) 11.5　　b) 18　　c) 9　　d) 4.76
2. 24
3. 308
4. 40
5. 52 kg

Exercise 7.4

1. £27.60　　2. £10.12　　3. £101.05　　4. £124,200　　5. £10.24
6. a) £109.75　　b) £96.58
7. 62.32 kg　　8. £2176　　9. £26.17　　10. 13.33%

Chapter 8 - Presentation of data

Exercise 8.1

1. a) Discrete　b) Continuous　c) Continuous　d) Discrete　e) Continuous　f) Discrete
2. Frequencies are: 1, 2, 2, 4, 5, 7, 5, 4, 4, 2, 1
3. Frequencies are: 8, 9, 10, 4, 3, 2, 0, 1, 1, 2
4. Frequencies are: 3, 1, 13, 6, 3, 2, 2, 2

Chapter 9 - Averaging Data

Exercise 9.1

1. a) 4.30m　　b) 4.20m　　c) 4.25m　　d) 0.75m
2. £1.84
3. 15½
4. £301 and £280
5. a) 212 hours　　b) 220 hours　　c) No mode　　d) Median
6. 68 marks and 41 marks
7. a) £509　　b) £459　　c) £640
8. 3.5 kg
9. a) 67 minutes　　b) 74 minutes
10. £1176

Chapter 10 - Perimeters and Areas

Exercise 10.1

1. a) 34 cm b) 44 in c) 36 ft d) 32 m
2. a) 78 cm b) 50 cm c) 52 cm
3. 124 cm
4. 1110 cm
5. 41 ft
6. a) 44 in b) 28 in
7. 164 cm

Exercise 10.2

1. a) 103.32 cm² b) 27.36 ft² c) 138.65 in²
   d) 43.68 m² e) 1399.56 m² f) 286.88 cm²
2. a) 6 cm b) 36 cm²
3. 9060 cm²
4. a) 140 cm² b) 46 cm² c) 77 cm²
5. a) 48 in² b) 72 in²
6. a) 3.7 m b) 17 m
7. 73.5 ft²
8. a) 3150 cm² b) 53 in² c) 80 ft²
9. a) 79 m² b) 123 m² c) 62 m² d) 264 m²
10. a) 29 cm b) 1189 cm²

Exercise 10.3

1. a) 36.72 cm² b) 59.04 cm² c) 211.12 m² d) 54.6 in²
2. 5.75 cm
3. 8.9 ft
4. a) 93 cm² b) 186 cm²

Exercise 10.4

1. a) 43.7 cm² b) 124.5 m² c) 34.56 ft² d) 18.33 cm² e) 93.33 m²
2. a) 9.1 cm b) 23 m c) 9.5 cm
3. a) 126.5 ft² b) 154 ft²

Exercise 10.5

1. a) 51.52 cm b) 77.91 m c) 30.47 ft d) 111.84 m
   e) 260.75 m f) 72.26 cm g) 56.86 cm h) 45.24 in
2. a) 132 ft b) 4.4 m c) 88 ft d) 1100 cm
   e) 88 in f) 264 cm
3. a) 4.3 cm b) 1.50 m c) 2.96 ft d) 2.05 in
4. a) 6.27 in b) 4.33 cm c) 2.71 m d) 17.92 cm
5. 392.7 cm
6. 18.46 cm
7. 11 in
8. 52.52 m
9. a) 207.34 cm b) 482.29 revs
10. a) 2ft b) 12.57 ft

Exercise 10.6

1. a) 154 cm²   b) 1386 ft²   c) 5544 mm²
2. a) 19.17 cm²   b) 1098.73 ft²   c) 141.04 m²
   d) 490.94 in²   e) 1676.60 cm²   f) 162.88 cm²
3. a) 3.07 cm   b) 3.37 ft   c) 4.18 cm   d) 1.68 m   e) 3.72 in   f) 2.34 m
4. a) 2.95 m   b) 18.56 m
5. a) 706.95 cm²   b) 314.2 cm²   c) 4713 cm²

Chapter 11 - Volumes

Exercise 11.1

1. a) 1800 cm³   b) 1689.6 m³   c) 302.328 cm³   d) 42.768 cm³
   e) 46.512 ft³   f) 1572.35 cm³   g) 35.03 m³
2. 103.823 cm³
3. 1954.08 cm³
4. 214.2 m³
5. 2601 in³
6. 2406.1 in³
7. a) 88704 cm³   b) 45969 cm³
8. 0.8096 m³

Exercise 11.2

1. a) 1256.8 cm³   b) 61498 mm³   c) 186157 cm³   d) 164088 cm³
2. 1944 cm³
3. 464 cm³
4. 20.72 cm
5. 0.7089 m³
6. a) 5221 cm³   b) 5.22 litres
7. a) 2827 cm³   b) 12.46 cm

Index

A
Addition, 15
Area, 65
Arithmetic mean, 61
Average, 61

C
Calculation
 order of, 18
Centigrade, 21
Circle, 75
Circumference, 72
Column, 6
Continuous variable, 55
Conversion
 of fractions, 26, 27
 of units, 19
Cube, 77
Cuboid, 77
Cylinder, 79

D
Denominator, 25
Directed numbers, 15
Discrete variable, 55
Division, 17

E
Equivalent fractions, 26

F
Fahrenheit, 21
Foreign exchange, 23
Formatting toolbar, 7
Fractions, 25
Frequency distribution, 54

G
Grouped distributions, 55

M
Maps, 37
Map scales, 37
Mean, 62
Median, 62
Menu, 7
Mode, 62
Multiplication, 17

N
Numerator, 25

O
Order of calculation, 18

P
Parallelogram, 66
 area of, 70
Percentages, 45
Perimeter, 65
Pie chart, 58
Plans, 37, 42
Presentation of data, 53
Proportional values, 31

Q
Quadrilateral, 65

R
Range, 63
Ration, 29
 simplification, 29
Raw data, 53
Rectangle, 68
Row, 6

S
Scales, 37
Spreadsheet, 5
Standard toolbar, 6
Subtraction, 15

T
Tally chart, 53
Three-dimensional representation, 44
Toolbar, 6
Traingle, 71

U
Units, 19

V
Variables, 55
Volume, 77